Lecture Notes in Mathematics

A collection of informal reports and seminars
Edited by A. Dold, Heidelberg and B. Eckmann, Zürich

T0239447

181

Frank DeMeyer
Colorado State University, Fort Collins, CO / USA

Edward Ingraham
Michigan State University, East Lansing, MI / USA

Separable Algebras Over Commutative Rings

Springer-Verlag
Berlin · Heidelberg · New York 1971

ISBN 3-540-05371-9 Springer-Verlag Berlin · Heidelberg · New York
ISBN 0-387-05371-9 Springer-Verlag New York · Heidelberg · Berlin

Offsetdruck: Julius Beltz, Weinheim/Bergstr.

PREFACE

This manuscript has its origin in courses given by the authors at Purdue University and Michigan State University. We were first introduced to this material by D. K. Harrison and we hope his influence is apparent. We also had access to notes prepared by G. J. Janusz for a course he gave at the University of Chicago.

We have laid the foundation for the study of the Brauer group of a commutative ring but have omitted the "spectral sequences" and the "method of descent" in the hope this will make the material more accessible. Cross-references are made as follows: Theorem 2.1 of Chapter 3 is referred to as Theorem 2.1 if the reference is in Chapter 3 and as Theorem 3.2.1 otherwise. We would like to thank the many students who have influenced and encouraged this publication, and Mrs. Glendora Milligan, who carefully typed the final form for reproduction.

September 1970

Frank DeMeyer
Colorado State University

Edward Ingraham
Michigan State University

CONTENTS

CHAPTER I

The purpose of this chapter is to present background material from the theory of rings which we will need in our consideration of separable algebras. The most important results of the chapter are the Morita Theorems (§3) and the theorem of the invariance of rank for projective modules over commutative rings with no idempotents but 0 and 1 (§4).

We begin with some fundamental definitions and conventions. Throughout we will be considering associative rings possessing an identity element 1. We will always suppose that subrings contain the identity of the overring and that a ring homomorphism from a ring R to a ring S takes the identity of R to the identity of S. Finally, all modules will be unitary. In what follows, "module" means "left module."

Suppose R is a commutative ring. By an R-algebra A we mean a ring A along with a ring homomorphism θ of R into the center of A. This induces a natural R-module structure on A by defining

$$r \cdot a = \theta(r)a \quad \text{for } r \text{ in } R,\ a \text{ in } A.$$

We then see that

$$(*) \qquad r\cdot(ab) = (r\cdot a)b = a(r\cdot b) \quad \text{for } r \text{ in } R,\ a, b \text{ in } A.$$

Conversely, if A is a ring which is also an R-module satisfying (*), then $r \to r\cdot 1$ is a ring homomorphism of R into the center of A and so A is an R-algebra. Accordingly we will write $R\cdot 1$ for the image of R under θ and we will use these two descriptions of an R-algebra interchangeably.

We describe an R-algebra as finitely generated or faithful if it is finitely generated or faithful when considered as an R-module. Thus A is a faithful R-algebra if and only if the homomorphism θ from R into the center of A is one-one. In this event $R \cong R\cdot 1$, so we identify R with $R\cdot 1$.

Let R be a ring. Suppose $\{M_i\}_{i\in I}$ is a family of R-modules, where I is some (not necessarily finite) indexing set. Let $\oplus \sum_{i\in I} M_i$ denote the set of all functions f from I into $\bigcup_i M_i$ such that f(i) is in M_i and f(i) = 0 for all but a finite set of indices i of I. Then $\oplus \sum M_i$ has a natural structure as an R-module given by

$$(f + g)(i) = f(i) + g(i)$$

and $$(r\cdot f)(i) = r\cdot f(i) \qquad \text{for } i \text{ in } I,\ r \text{ in } R,$$

f, g in $\oplus \sum M_i$. We call this R-module $\oplus \sum M_i$ the direct sum of the family $\{M_i\}_{i\in I}$. In the event I = {1, 2,..., n}, we sometimes write $\oplus \sum M_i$ as $M_1 \oplus \cdots \oplus M_n$.

Finally, we speak of a submodule N of an R-module M as being a direct summand of M if M contains a submodule L such that N + L = M and N ∩ L = (0). In this case, it is apparent that M ≅ N ⊕ L.

§ 1. Progenerator Modules

A. Free and projective modules. For any indexing set I, we denote by R^I the module $\oplus \sum_{i\in I} R_i$ where each R_i equals R. If I = {1,..., n}, we write $R^{(n)}$ for R^I. An R-module M is called a free R-module if M is isomorphic as an R-module to R^I for some set I. If we define b_i in R^I by $b_i(j) = \delta_{i,j}$, i,j in I,(*) then we see that

1.) every element of R^I has a representation of the form $\sum_{i\in I'} r_i b_i$ where r_i is from R and I′ is some finite subset of I; and,

2.) for any finite subset I′ of I, $\sum_{i\in I'} r_i b_i = 0$ implies $r_i = 0$ for all i in I′.

On the other hand, if M is an R-module and $\{b_i\}_{i\in I}$ is some subset of M satisfying 1.) and 2.), the map $\psi: R^I \to M$ defined by $\psi(f) = \sum_{i\in I} f(i)b_i$ can be seen to be an R-module isomorphism, whence M is free. It follows that

Lemma 1.1: An R-module M is free if and only if there exists

(*) Throughout these notes, when a, b are elements of some set S, we define $\delta_{a,b}$ to be 1 if a = b and 0 if a ≠ b.

$\{b_i\}_{i\in I} \subset M$ (I some indexing set) satisfying 1.) and 2.) above. The set $\{b_i\}_{i\in I}$ is called a <u>basis</u> for M.

When R is a commutative ring, it can be shown that if a free R-module has a finite basis consisting of n elements, then any other basis consists of n elements. This integer n is called the <u>rank</u> of M.

<u>Proposition 1.2:</u> The following properties of an R-module M are equivalent:

a.) M is isomorphic as an R-module to a direct summand of some free R-module.

b.) Every short exact sequence of R-modules

$$0 \to L \to N \overset{\varphi}{\to} M \to 0$$

splits; that is, there exists an R-module homomorphism ρ: $M \to N$ such that $\varphi\rho = i_M$, where i_M denotes the identity map on M.

c.) For any diagram of R-modules and R-homomorphisms

$$\begin{array}{ccc} & M & \\ & \downarrow \mu & \\ N \overset{\alpha}{\to} & N' & \to 0 \end{array}$$

with the bottom row exact, there exists an R-homomorphism β: $M \to N$ with $\alpha\beta = \mu$.

An R-module satisfying the equivalent conditions of Proposition 1.2 is called <u>projective</u>. An R-algebra A is called <u>projective</u> if it is projective as an R-module.

We isolate another equivalent condition for projectivity in the following lemma, which we call the Dual Basis Lemma.

<u>Lemma 1.3</u>: Let M be an R-module. Then M is projective if and only if there exists $\{m_i\}_{i\in I} \subset M$ and $\{f_i\}_{i\in I} \subset \mathrm{Hom}_R(M,R)$ (I some indexing set) such that

a.) for every m in M, $f_i(m) = 0$ for all but a finite subset of i in I; and,

b.) for every m in M, $\sum_{i \in I} f_i(m)m_i = m$.

Moreover, I is a finite set if and only if M is finitely generated. The collection $\{m_i, f_i\}$ is called a <u>dual basis</u> for M.

<u>Proof</u>: Supposing M is projective, there exists a set I such that M is isomorphic as a left R-module to a direct summand of the free R-module R^I. Equivalently, there exist R-module maps $\varphi: M \to R^I$ and $\pi: R^I \to M$ such that $\pi\varphi = i_M$. Thinking of R^I as a set of functions from I to R, let π_i denote the function from R^I to R given by $\pi_i(f) = f(i)$, for all f in R^I. Then for any f in R^I, we have $\sum_{i \in I} \pi_i(f)b_i = f$ since $[\sum_i \pi_i(f)b_i](j) = \pi_j(f) = f(j)$. (Recall that b_i is that element of R^I defined by $b_i(j) = \delta_{i,j}$.) Thus $\{\pi_i, b_i\}$ forms a dual basis for R^I. Now set $m_i = \pi(b_i)$ and $f_i = \pi_i\varphi$. Clearly $f_i(m) = 0$ for almost all i in I and

$$\sum_{i \in I} f_i(m)m_i = \sum_{i \in I} \pi_i(\varphi(m))\pi(b_i) = \pi(\sum_{i \in I} \pi_i(\varphi(m))b_i) = \pi(\varphi(m)) = m,$$

for every m in M. Thus $\{f_i, m_i\}$ forms a dual basis for M.

Conversely, if $\{f_i, m_i\}$ is a dual basis for M, define $\varphi: M \to R^I$ by $\varphi(m)(i) = f_i(m)$ and $\pi: R^I \to M$ by $\pi(f) = \sum_{i \in I} f(i)m_i$. One can easily check that φ and π are R-module homomorphisms and

$$\pi(\varphi(m)) = \sum_{i \in I} f_i(m)m_i = m .$$

Thus $\pi\varphi = i_M$, whence M is isomorphic to a direct summand of R^I and is therefore projective.

Suppose R and S are (possibly non-commutative) rings and $\theta: R \to S$ is a ring homomorphism. Then S becomes an R-module under the operation $r \cdot s = \theta(r)s$. This naturally induces an R-module structure on any S-module. This setting typically arises when R is commutative and S is an R-algebra or when R is a subring of S and θ the identity.

As a typical application of the Dual Basis Lemma we prove

<u>Proposition 1.4</u>: (Transitivity of Projective Modules) Let R and S be rings and $\theta:R \to S$ a ring homomorphism such that S is projective when considered as an R-module. Then any projective S-module M is projective as an R-module. Furthermore, if M is finitely generated over S and S is finitely generated over R, M will be finitely generated over R.

<u>Proof</u>: By the Dual Basis Lemma (1.3), there exist $\{f_i\}_{i\in I} \subset \text{Hom}_S(M,S)$ and $\{m_i\}_{i\in I} \subset M$ with $f_i(m) = 0$ for almost all i in I and $\sum_{i\in I} f_i(m)m_i = m$, for every m in M. Similarly there exist $\{g_j\}_{j\in J} \subset \text{Hom}_R(S,R)$ and $\{s_j\}_{j\in J} \subseteq S$ with $g_j(s) = 0$ for almost all j in J and $\sum_{j\in J} g_j(s)s_j = s$, for every s in S. Then g_jf_i is in $\text{Hom}_R(M,R)$ and s_jm_i is in M with $g_jf_i(m) = 0$ for almost all (i,j) in I $\times$ J and

$$\sum_{i,j} g_j(f_i(m))s_jm_i = \sum_i \left(\sum_j g_j(f_i(m))s_j\right)m_i = \sum_i f_i(m)m_i = m .$$

Thus M is R-projective. The conclusion concerning finite generation is obvious.

<u>B. Generator modules</u>. For any R-module M, consider the subset $\mathfrak{T}_R(M)$ of R consisting of the elements of the form $\sum_i f_i(m_i)$ where the f_i are from $\text{Hom}_R(M,R)$ and the m_i are from M. Since $\text{Hom}_R(M,R)$ is a right R-module under the operation $(f\cdot r)(m) = f(m)r$, we see that $r(\sum_i f_i(m_i)) = \sum_i f_i(rm_i)$ is in $\mathfrak{T}_R(M)$ and $(\sum_i f_i(m_i))r = \sum_i (f_i\cdot r)(m_i)$ is in $\mathfrak{T}_R(M)$. It follows that $\mathfrak{T}_R(M)$ is a two-sided ideal of R, called the <u>trace ideal</u> of M. We call an R-module M an <u>R-generator</u> if $\mathfrak{T}_R(M) = R$. Therefore M is an R-generator if and only if there exist $f_1,\dots, f_n$ in $\text{Hom}_R(M,R)$ and $m_1,\dots, m_n$ in M with $\sum_i f_i(m_i) = 1$.

<u>Proposition 1.5</u>: (Transitivity of Generators) Let R and S be rings and θ: R $\to$ S a ring homomorphism such that S is a generator when considered as an R-module. Then any S-generator M is an R-generator.

Proof: There exist $f_1, \ldots, f_n$ in $\mathrm{Hom}_S(M,S)$ and $m_1, \ldots, m_n$ in M with $\sum_i f_i(m_i) = 1$. Similarly there exist $g_1, \ldots, g_m$ in $\mathrm{Hom}_R(S,R)$ and $s_1, \ldots, s_m$ in S with $\sum_j g_j(s_j) = 1$. Then $g_j f_i$ is in $\mathrm{Hom}_R(M,R)$ and $s_j m_i$ is in M with

$$\sum_{i,j} g_j f_i(s_j m_i) = \sum_{i,j} g_j(s_j \cdot f_i(m_i)) = \sum_j g_j(s_j \cdot \sum_i f_i(m_i)) = 1,$$

so M is an R-generator.

We call an R-module M an R-progenerator if M is finitely generated, projective and a generator over R. Modules of this type will play an important part in what follows. We observe that as a result of the transitivity of projectivity (1.4), finite generation (1.4), and being a generator (1.5), we have

Proposition 1.6: (Transitivity of Progenerators) Let R and S be rings and θ: R $\to$ S a ring homomorphism such that S is an R-progenerator when considered as an R-module. Then any S-progenerator M is an R-progenerator.

It now seems an opportune moment to prove one of the standard tools of ring theory, a form of the so-called Nakayama's Lemma. For any ideal $\mathfrak{a}$ of a ring R and any R-module M, $\mathfrak{a}M$ means the set of all finite sums of the form $\sum \alpha_i m_i$ where α_i comes from $\mathfrak{a}$ and m_i comes from M. Define the annihilator of M in R by $\mathrm{annih}_R(M) = \{r \in R \mid r \cdot m = 0$ for all m in M$\}$.

Lemma 1.7: (Generalized Nakayama Lemma) Let R be a commutative ring and M a finitely generated R-module. An ideal $\mathfrak{a}$ of R has the property that $\mathfrak{a}M = M$ if and only if $\mathfrak{a} + \mathrm{annih}_R(M) = R$.

Proof: If $\mathfrak{a} + \mathrm{annih}_R(M) = R$, we can write $1 = \alpha + \beta$ for some α in $\mathfrak{a}$, β in $\mathrm{annih}_R(M)$, so that for every m in M, $m = 1m = \alpha m + \beta m = \alpha m$, so $\mathfrak{a}M = M$.

To prove the converse, we suppose $M = Rm_1 + \cdots + Rm_n$ and we proceed by induction. Set $M_i = Rm_i + \cdots + Rm_n$, $i = 1,2,\ldots, n$ and $M_{n+1} = 0$. We will show that for every $i = 1, 2, \ldots, n+1$, there exists α_i in $\mathfrak{a}$ with $(1 - \alpha_i)M \subset M_i$. Clearly we may choose $\alpha_1 = 0$. Suppose we have α_i with $(1 - \alpha_i)M \subset M_i$. Then $(1 - \alpha_i)M = (1 - \alpha_i)\mathfrak{a}M$ $= \mathfrak{a}(1 - \alpha_i)M \subset \mathfrak{a}M_i$, so $(1 - \alpha_i)m_i = \sum_{j=1}^{n} \alpha_{ij}m_j$ or $(1 - \alpha_i - \alpha_{ii})m_i$ is in M_{i+1}, where the α_{ij} come from $\mathfrak{a}$. Then $(1-\alpha_i)(1-\alpha_i-\alpha_{ii})M \subset (1-\alpha_i-\alpha_{ii})M_i$ $\subset M_{i+1}$, so $[1-(2\alpha_i+\alpha_{ii}-\alpha_i^2-\alpha_i\alpha_{ii})]M \subset M_{i+1}$. We complete the induction step by setting $\alpha_{i+1} = 2\alpha_i + \alpha_{ii} - \alpha_i^2 - \alpha_i\alpha_{ii}$. Then $1 - \alpha_{n+1}$ is in $\operatorname{annih}_R(M)$ and we are done.

<u>Corollary 1.8</u>: Let M be a finitely generated module over a commutative ring R. If $\mathfrak{m}M = M$ for every maximal ideal $\mathfrak{m}$ of R, then $M = 0$.

<u>Proof</u>: $M = 0$ if and only if $\operatorname{annih}_R(M) = R$, but if $\operatorname{annih}_R(M) \neq R$, there is some maximal ideal $\mathfrak{m}$ with $\operatorname{annih}_R(M) \subset \mathfrak{m}$, whence $\operatorname{annih}_R(M) + \mathfrak{m} \neq R$, a contradiction.

It is often very difficult to tell whether a finitely generated, projective module is a generator. However, in the case of a commutative ring, there is an easy criterion.

<u>Proposition 1.9</u>: Let R be a commutative ring and M a finitely generated, projective R-module. Then $\mathfrak{T}_R(M) \oplus \operatorname{annih}_R(M) = R$.

<u>Proof</u>: Since M is finitely generated and projective, there exists a dual basis $f_1, \ldots, f_n$ in $\operatorname{Hom}_R(M,R)$ and $m_1, \ldots, m_n$ in M. Then $m = \sum_i f_i(m)m_i$ for every m in M and $f_i(m)$ is in $\mathfrak{T}_R(M)$, so $\mathfrak{T}_R(M)M = M$. Thus by (1.7) $\mathfrak{T}_R(M) + \operatorname{annih}_R(M) = R$. But $\mathfrak{T}_R(M)\cdot\operatorname{annih}_R(M) = 0$, since for any α in $\operatorname{annih}_R(M)$, f in $\operatorname{Hom}_R(M,R)$ and m in M we have $\alpha f(m) = f(\alpha m) = 0$. It follows that for any x in $\mathfrak{T}_R(M) \cap \operatorname{annih}_R(M)$, $x = 1x$ $= \alpha x + \beta x = 0$ where $1 = \alpha + \beta$ with β in $\mathfrak{T}_R(M)$ and α in $\operatorname{annih}_R(M)$. Thus $\mathfrak{T}_R(M) \cap \operatorname{annih}_R(M) = (0)$ and $\mathfrak{T}_R(M) \oplus \operatorname{annih}_R(M) = R$.

Corollary 1.10: If R is a commutative ring, an R-module M is an R-progenerator if and only if M is finitely generated, projective and faithful.

Corollary 1.11: If R is a commutative ring with no idempotents but 0 and 1, every finitely generated, projective non-zero R-module is faithful and therefore a progenerator.

§ 2. Categories and Functors of Modules

In these notes we will consider for the most part categories whose objects are modules. We therefore omit the general definition of category and restrict our concern to the collection of all modules over some ring. For a ring R, we will let ${}_R\mathfrak{M}$ denote the category of left R-modules, that is, the collection of all left R-modules along with, for any two R-modules M and N, the set $\mathrm{Hom}_R(M,N)$ of all R-module homomorphisms from M to N. Similarly $\mathfrak{M}_R$ will denote the category of right R-modules and R-homomorphisms.

For any ring R, we let R° denote the ring whose underlying abelian group is the same as that of R but whose multiplication * is given by $a*b = ba$ where ba is the product in R. R° is called the opposite ring of R. It is clear that a left R-module M can be given the structure of a right R°-module by defining $m \cdot' r = r \cdot m$ (r in R, m in M) where $r \cdot m$ is the scalar product of r and m. Thus in particular if R is commutative, $R = R^\circ$ and any one-sided R-module can be considered as both a right and a left R-module. Consequently, when R is commutative, ${}_R\mathfrak{M} = \mathfrak{M}_R$.

By a (covariant) functor from a category of modules $\mathfrak{C}$ to a category of modules $\mathfrak{D}$, we mean a correspondence $\mathfrak{F}$ which associates to every module M in $\mathfrak{C}$ a module $\mathfrak{F}(M)$ in $\mathfrak{D}$ and which associates to every homomorphism $f: M \to N$ in $\mathfrak{C}$ a homomorphism $\mathfrak{F}(f): \mathfrak{F}(M) \to \mathfrak{F}(N)$ such that for every M in $\mathfrak{C}$

a.) $\mathfrak{F}(i_M) = i_{\mathfrak{F}(M)}$ where $i_{()}$ is the identity homomorphism; and,

b.) if f, g are homomorphisms in $\mathfrak{C}$ such that fg is defined, then $\mathfrak{F}(fg) = \mathfrak{F}(f)\mathfrak{F}(g)$.

A functor $\mathfrak{F}$ from a category $\mathfrak{C}$ to a category $\mathfrak{D}$ is called <u>left exact</u> if for every exact sequence of modules and maps

$$(*) \qquad 0 \to L \to M \to N \to 0$$

in $\mathfrak{C}$,

$$0 \to \mathfrak{F}(L) \to \mathfrak{F}(M) \to \mathfrak{F}(N)$$

is exact in $\mathfrak{D}$. Similarly $\mathfrak{F}$ is called <u>right exact</u> if (*) always gives rise to

$$\mathfrak{F}(L) \to \mathfrak{F}(M) \to \mathfrak{F}(N) \to 0$$

exact in $\mathfrak{D}$. A functor that is both left and right exact is called <u>exact</u>.

The two functors we will be most concerned with are the tensor product and the set of all module homomorphisms from some fixed module to another. We now review the fundamental properties of these concepts and give some results we will need.

<u>A. Tensor products</u>. We assume the reader has some familiarity with the idea of the tensor product of modules. We review the definition and list the fundamental properties.

Recall that given a right R-module M and a left R-module N, we form their tensor product $M \otimes_R N$ by first considering the free abelian group $Z^{M \times N}$ indexed by the cartesian product $M \times N$ (where Z denotes the rational integers) and then factoring out the subgroup $\mathfrak{R}$ generated by all elements of the form

$$b_{(m+m',n)} - b_{(m,n)} - b_{(m',n)},\ b_{(m,n+n')} - b_{(m,n)} - b_{(m,n')},$$

and $b_{(m\cdot r,n)} - b_{(m,r\cdot n)}$, for all m, m′ in M, n, n′ in N and r in R .

Here $b_{(m,n)}$ denotes the basis element of $Z^{M\times N}$ determined by (m,n). We will denote the coset $b_{(m,n)} + \mathfrak{R}$ by $m\otimes n$. As a result of this definition we have the identities $(m+m')\otimes n = (m\otimes n) + (m'\otimes n)$, $m\otimes(n+n') = (m\otimes n) + (m\otimes n')$ and $(m\cdot r)\otimes n = m\otimes(r\cdot n)$, for all m,m' in M, n,n' in N and r in R. Notice that a typical element of $M \otimes_R N$ looks like $\sum_{i=1}^{k} m_i \otimes n_i$ and cannot necessarily be represented by a monomial $m \otimes n$.

Let R and S be rings. We call an abelian group M a <u>left R-left S bimodule</u> if M is a left R-module as well as a left S-module such that $r(sm) = s(rm)$ for all r in R, s in S and m in M. We let ${}_{R-S}\mathfrak{M}$ stand for the category of left R-left S bimodules, where the morphisms of the category are the maps which are both R- and S-homomorphisms. We can similarly define the category ${}_R\mathfrak{M}_S$ of left R-right S bimodules and the category $\mathfrak{M}_{R-S}$ of right R-right S bimodules.

As matters stand, $M \otimes_R N$ has only the structure of an abelian group. However, under certain circumstances, $M \otimes_R N$ can be given an additional module structure. This can be illustrated by the following: suppose $N \in {}_R\mathfrak{M}_S$, that is, N is a left R-right S bimodule. Then $M \otimes_R N$ becomes a right S-module under the operation $(\sum_i m_i \otimes n_i)s = \sum_i m_i \otimes (n_i s)$, this operation being well-defined because of the bimodule structure on N. The same sort of thing can be done for $M \in {}_S\mathfrak{M}_R$, $M \in \mathfrak{M}_{R-S}$ and $N \in {}_{R-S}\mathfrak{M}$. In particular we see that if R is commutative, every R-module can be viewed as an R-R bimodule (in all possible ways), so that $M \otimes_R N$ has the structure of an R-module.

The following are well-known properties of the tensor product. Their proofs may be found in many sources, among which are C. W. Curtis and I. Reiner "Representation Theory of Finite Groups and Associative Algebras," [G], or, D. G. Northcott "An Introduction to Homological Algebra." [I]

1.) For any ring R and any right R-module M, left R-module N, $M \otimes_R N \cong N \otimes_{R^\circ} M$ under the map $m\otimes n \to n\otimes m$. Therefore if R is commutative,

$M \otimes_R N \cong N \otimes_R M$ as R-modules. *)

2.) (Associativity) For any rings R and S and $L \in \mathfrak{M}_R$, $M \in {}_R\mathfrak{M}_S$, and $N \in {}_S\mathfrak{M}$, we have $(L \otimes_R M) \otimes_S N \cong L \otimes_R (M \otimes_S N)$ under the map $(\ell \otimes m) \otimes n \to \ell \otimes (m \otimes n)$, where $L \otimes_R M$ is an S-module and $M \otimes_S N$ is an R-module by virtue of the bimodule structure on M.

3.) For any ring R and any $M \in {}_R\mathfrak{M}$, $R \otimes_R M \cong M$ under the map $r \otimes m \to rm$. Similarly for $M \in \mathfrak{M}_R$, $M \otimes_R R \cong M$.

4.) For $M, M' \in \mathfrak{M}_R$, $N, N' \in {}_R\mathfrak{M}$, $f \in \mathrm{Hom}_R(M,M')$ and $g \in \mathrm{Hom}_R(N,N')$, there is a well-defined homomorphism $f \otimes g$ from $M \otimes_R N$ to $M' \otimes_R N'$ given by $f \otimes g\ (m \otimes n) = f(m) \otimes g(n)$.

If we fix a right R-module M, then $M \otimes_R (\ )$ can be considered a functor from ${}_R\mathfrak{M}$ to ${}_Z\mathfrak{M}$ which takes $N \in {}_R\mathfrak{M}$ to $M \otimes_R N$ and any $f \in \mathrm{Hom}_R(N,N')$ to $i_M \otimes f$ in $\mathrm{Hom}_Z(M \otimes_R N, M \otimes_R N')$. Similarly $(\ ) \otimes_R N$ is a functor from $\mathfrak{M}_R$ to ${}_Z\mathfrak{M}$.

5.) For any exact sequence $O \to N_1 \xrightarrow{\alpha} N_2 \xrightarrow{\beta} N_3 \to O$ of left R-modules,

$$M \otimes_R N_1 \xrightarrow{i\otimes\alpha} M \otimes_R N_2 \xrightarrow{i\otimes\beta} M \otimes_R N_3 \longrightarrow O$$

is exact for any right R-module M. Thus $M \otimes_R (\ )$ is a right exact functor. Similarly $(\ ) \otimes_R N$ is a right exact functor. In the event that tensoring by a module is exact, that is, it is left exact as well as right exact, we say the module is <u>flat</u>.

6.) Let $\{M_i\}_{i \in I}$ be a collection of right R-modules and let $\{N_j\}_{j \in J}$

[*) We will often give a definition involving elements of a tensor product in terms of a single element $m\otimes n$. It will always be understood that such a definition is to be extended to an arbitrary element $\sum m_i \otimes n_i$ of the tensor product by linearity. Consequently, in this case, $m\otimes n \to n\otimes m$ induces $\sum m_i\otimes n_i \to \sum n_i \otimes m_i$.

Furthermore, we should caution that a candidate for a mapping on a tensor product, being "defined" on a factor group, is in danger of not being well-defined. We will rarely if ever check to see that a map we suggest is well-defined but the conscientious reader should verify that a map f "defined" on $M \otimes_R N$ is R-bilinear, that is, $f((m+m')\otimes n) = f(m\otimes n) + f(m'\otimes n)$, $f(m\otimes(n+n')) = f(m\otimes n) + f(m\otimes n')$ and $f((m\cdot r)\otimes n) = f(m\otimes(r\cdot n))$.]

be a collection of left R-modules. Then we have $(\oplus \sum_{i \in I} M_i) \otimes_R (\oplus \sum_{j \in J} N_j) \cong \oplus \sum_{i,j} (M_i \otimes_R N_j)$ under the map $f \otimes g \to h$ where $h((i,j)) = f(i) \otimes g(j)$. Thus we say that tensor products distribute over direct sums.

7.) It follows easily from 6.), 3.) and the characterization of projective modules as direct summands of free modules that every projective R-module is flat.

8.) (Universal Mapping Property) For a right R-module M and a left R-module N, $M \otimes_R N$ has the property that if G is any abelian group and f is any map from $M \times N$ to G satisfying

i.) $f(m + m', n) = f(m,n) + f(m', n)$;

ii.) $f(m,n + n') = f(m,n) + f(m, n')$; and,

iii.) $f(mr, n) = f(m, rn)$,

for all m, m′ in M, n, n′ in N and r in R, then there exists a unique homomorphism $f^*: M \otimes_R N \to G$ with $f^*(m \otimes n) = f(m,n)$, for all m in M, n in N. It follows easily from this that any other abelian group satisfying the universal mapping property of $M \otimes_R N$ is isomorphic to $M \otimes_R N$; that is, if H is an abelian group and h is a map from $M \times N$ to H satisfying i.), ii.) and iii.) such that for any f as given above, there exists a unique homomorphism $f^*: H \to G$ with $f^*h = f$, then $H \cong M \otimes_R N$.

In addition to these properties, we need a few more specialized results on the behavior of modules under $\otimes$.

<u>Proposition 2.1</u>: Let R be a commutative ring and let A and B be R-algebras. If M is a right A-module, then $M \otimes_R B$ is projective (resp. a generator) over $A \otimes_R B$ if M is projective (resp. a generator) over A.

<u>Proof</u>: Assuming M to be A-projective, there exists a dual basis $\{f_i \in \mathrm{Hom}_A(M,A), m_i \in M\}$. One easily verifies that $\{f_i \otimes i_B, m_i \otimes 1\}$ forms a dual basis for $M \otimes_R B$ over $A \otimes_R B$. Similarly if M is an A-generator, there exist $f_i \in \mathrm{Hom}_A(M,A)$ and $m_i \in M$ with $\sum f_i(m_i) = 1$.

Then $\sum (f_i \otimes i_B)(m_i \otimes 1) = 1 \otimes 1$, so $M \otimes_R B$ is an $A \otimes_R B$-generator.

Corollary 2.2: If M is projective (resp. a generator) over R, then for any R-algebra B, $M \otimes_R B$ is projective (resp. a generator) over B.

A similar proof, which we leave as an exercise, yields the following

Proposition 2.3: Let R be a ring, M and N R-modules. Then

a.) $M \otimes_R N$ is finitely generated over R if both M and N are.

b.) $M \otimes_R N$ is R-projective if both M and N are.

c.) $M \otimes_R N$ is an R-generator if both M and N are.

d.) $M \otimes_R N$ is an R-progenerator if both M and N are.

B. Hom. As above, for R-modules M and N we let $\mathrm{Hom}_R(M,N)$ denote the set of all functions f from M to N such that

i.) $f(m + m') = f(m) + f(m')$, and

ii.) $f(r \cdot m) = r \cdot f(m)$, for all $r \in R$, $m, m' \in M$.

$\mathrm{Hom}_R(M,N)$ has the structure of an abelian group under the operation $(f + g)(m) = f(m) + g(m)$, for $f, g \in \mathrm{Hom}_R(M,N)$ and $m \in M$.

If, in addition, M or N is a bimodule, for example $M \in {}_R\mathfrak{M}_S$, then $\mathrm{Hom}_R(M,N)$ can be endowed with an S-module structure by a suitable operation which, in the case of $M \in {}_R\mathfrak{M}_S$, is $(s \cdot f)(m) = f(m \cdot s)$ for all $s \in S$, $m \in M$ and $f \in \mathrm{Hom}_R(M,N)$. We leave it to the reader to resolve in his own mind how to define the S-operation in various other situations such as $M \in {}_{R-S}\mathfrak{M}$, $N \in {}_R\mathfrak{M}_S$, etc. However, we remark that when R is a commutative ring, every R-module can be considered as an R-R-bimodule so that, as with the tensor product, $\mathrm{Hom}_R(M,N)$ becomes an R-module.

Further, when M = N, functional composition of elements of $\mathrm{Hom}_R(M,N)$ serves as a multiplication under which $\mathrm{Hom}_R(M,M)$ becomes a ring. Thus, when R is commutative, $\mathrm{Hom}_R(M,M)$ has the structure of an

R-algebra.

For any fixed R-module M and any homomorphism f from an R-module N to an R-module N', we have a mapping $\mathrm{Hom}_R(i_M, f)$ from $\mathrm{Hom}_R(M,N)$ to $\mathrm{Hom}_R(M,N')$ given by $\mathrm{Hom}_R(i_M,f)(g) = fgi_M = fg$. With this in mind it is easy to check that $\mathrm{Hom}_R(M, -)$ can be viewed as a functor from ${}_R\mathfrak{M}$ to ${}_Z\mathfrak{M}$. The following fundamental properties of Hom are easily verified.

1.) $\mathrm{Hom}_R(M, -)$ is a left exact functor which by Proposition 1.2 is exact if and only if M is R-projective.

2.) For $M_1,\ldots, M_n, N_1,\ldots, N_k$ R-modules, we have $\mathrm{Hom}_R(\oplus \sum_{i=1}^{n} M_i , \oplus \sum_{j=1}^{k} N_j) \cong \oplus \sum_{i,j}^{n,k} \mathrm{Hom}_R(M_i,N_j)$ where the map is given by $f \to g$ with $g((i,j))(m_i) = \pi_j f(0,\ldots,0,m_i,0,\ldots,0)$ where m_i is in the i-th spot of the n-tuple and π_j is the projection of N onto N_j.

3.) $\mathrm{Hom}_R(R,M) \cong M$ under $f \to f(1)$.

We next give some isomorphisms involving the relationship between Hom and the tensor product. We informally speak of these as the Hom-Tensor Relations.

Hom-Tensor Relation 2.4: Let R be a commutative ring and let A and B be R-algebras. Let M be a finitely generated and projective A-module and let N be a finitely generated and projective B-module. Then, for any A-module M' and any B-module N', the mapping ψ: $\mathrm{Hom}_A(M,M') \otimes_R \mathrm{Hom}_B(N,N') \to \mathrm{Hom}_{A\otimes_R B}(M\otimes_R N, M'\otimes_R N')$ induced by $\psi(f\otimes g)(m\otimes n) = f(m)\otimes g(n)$ is an R-module isomorphism. If $M = M'$ and $N = N'$, ψ is an algebra map as well. (By an algebra map, we mean a ring homomorphism that is simultaneously a module homomorphism.)

Proof: That ψ is a well-defined R-module homomorphism which preserves multiplication when $M = M'$ and $N = N'$ is easily checked. Therefore we need only show that ψ is one-one and onto.

Case 1: Suppose $M = A$ and $N = B$. Then property 3.) above yields that ψ is an isomorphism.

Case 2: Suppose $M = A^{(m)}$ and $N = B^{(n)}$, that is, M and N are free of finite rank. Then by the distributivity of Hom and ⊗ over finite direct sums, we have

$\mathrm{Hom}_A(A^{(m)},M') \otimes_R \mathrm{Hom}_B(B^{(n)}, N') \cong \mathrm{Hom}_A(A,M')^{(m)} \otimes_R \mathrm{Hom}_B(B,N')^{(n)}$ $\cong M'^{(m)} \otimes_R N'^{(n)} \cong \oplus\Sigma\, M' \otimes_R N'$ (mn-times) $\cong \oplus\Sigma\, \mathrm{Hom}_{A\otimes_R B}(A\otimes_R B, M'\otimes_R N')$ (mn-times) $\cong \mathrm{Hom}_{A\otimes_R B}(A^{(m)}\otimes_R B^{(n)}, M' \otimes_R N')$ where the composition of these isomorphisms is ψ.

General case: Suppose M and N are finitely generated and projective. Then there exist integers m, n, an A-module L and a B-module K such that $M \oplus L \cong A^{(m)}$ and $N \oplus K \cong B^{(n)}$. It follows that there is an R-submodule H of $\mathrm{Hom}_A(A^{(m)}, M') \otimes_R \mathrm{Hom}_B(B^{(n)}, N')$ and an R-submodule H' of $\mathrm{Hom}_{A\otimes_R B}(A^{(m)}\otimes_R B^{(n)}, M' \otimes_R N')$ such that $[\mathrm{Hom}_A(M,M') \otimes_R \mathrm{Hom}_B(N,N')] \oplus H \cong \mathrm{Hom}_A(A^{(m)}, M') \otimes_R \mathrm{Hom}_B(B^{(n)}, N')$ and $\mathrm{Hom}_{A\otimes_R B}(M\otimes_R N, M'\otimes_R N') \oplus H' \cong \mathrm{Hom}_{A\otimes_R B}(A^{(m)}\otimes_R B^{(n)}, M' \otimes_R N')$ in such a fashion that the isomorphism established in Case 2 maps H into H' and equals ψ on $\mathrm{Hom}_A(M,M') \otimes_R \mathrm{Hom}_B(N,N')$. We can infer from this that ψ is one-one and onto.

__Corollary 2.5__: For any R-algebra A and any finitely generated, projective R-module N, $A \otimes_R \mathrm{Hom}_R(N,N') \cong \mathrm{Hom}_A(A\otimes_R N, A\otimes_R N')$ for any R-module N'.

__Proof__: Set $M = M' = A$ and $B = R$.

__Corollary 2.6__: If M and N are finitely generated, projective R-modules, then $\mathrm{Hom}_R(M,M) \otimes_R \mathrm{Hom}_R(N,N) \cong \mathrm{Hom}_R(M\otimes_R N, M\otimes_R N)$ as R-algebras.

__Proof__: Set $A = B = R$ and $M = M'$, $N = N'$.

__Hom-Tensor Relation 2.7__: Let A and B be rings. Suppose L is a finitely generated, projective A-module, $M \in {}_A\mathfrak{M}_B$ and $N \in {}_B\mathfrak{M}$. Then the map

$$\psi\colon \mathrm{Hom}_A(L,M)\otimes_B N \to \mathrm{Hom}_A(L, M\otimes_B N)$$

induced by $[\psi(f\otimes n)](\ell) = f(\ell)\otimes n$ is a group isomorphism.

Proof: The proof is analogous to that of Hom-Tensor Relation 2.4; that is, the relation clearly holds for $L = A$, can easily be extended to $L = A^{(n)}$ by distributivity of Hom and $\otimes$ over finite direct sums, and finally is verified for the general case by considering L as a direct summand of some $A^{(n)}$.

§ 3. The Morita Theorems

Let $\mathfrak{C}$ and $\mathfrak{D}$ be categories of modules and suppose we have two functors $\mathfrak{F}$ and $\mathfrak{F}'$ from $\mathfrak{C}$ to $\mathfrak{D}$. We say that $\mathfrak{F}$ and $\mathfrak{F}'$ are naturally equivalent if for every module M in $\mathfrak{C}$ there is an isomorphism φ_M in $\mathrm{Hom}_{\mathfrak{D}}(\mathfrak{F}(M), \mathfrak{F}'(M))$ such that, for every pair of modules M and N in $\mathfrak{C}$ and any $f \in \mathrm{Hom}_{\mathfrak{C}}(M,N)$, the diagram

$$\begin{array}{ccc} \mathfrak{F}(M) & \xrightarrow{\mathfrak{F}(f)} & \mathfrak{F}(N) \\ \downarrow{\scriptstyle \varphi_M} & & \downarrow{\scriptstyle \varphi_N} \\ \mathfrak{F}'(M) & \xrightarrow{\mathfrak{F}'(f)} & \mathfrak{F}'(N) \end{array}$$

commutes. We denote by $I_{\mathfrak{C}}$ the identity functor on the category $\mathfrak{C}$ defined by $I_{\mathfrak{C}}(M) = M$ and $I_{\mathfrak{C}}(f) = f$, for modules M and maps f. Then we say two categories $\mathfrak{C}$ and $\mathfrak{D}$ are equivalent ($\mathfrak{C} \sim \mathfrak{D}$) if there is a functor $\mathfrak{F}: \mathfrak{C} \to \mathfrak{D}$ and a functor $\mathfrak{G}: \mathfrak{D} \to \mathfrak{C}$ such that $\mathfrak{F}\circ\mathfrak{G}$ is naturally equivalent to $I_{\mathfrak{D}}$ and $\mathfrak{G}\circ\mathfrak{F}$ is naturally equivalent to $I_{\mathfrak{C}}$. $\mathfrak{F}$ and $\mathfrak{G}$ are then referred to as inverse equivalences.

In this section we show that $\mathfrak{M}_R$ and ${}_S\mathfrak{M}$ are equivalent categories when S is chosen as the endomorphism ring of some R-progenerator. Our interest in this equivalence arises from the fact that a considerable amount of information concerning the relationship of the modules in the two categories can be obtained just from this category equivalence. This equivalence and its consequences are known as the Morita theorems and we speak of R and S as being Morita equivalent.

The only result of a general categorical nature that we need is the following

<u>Proposition 3.1</u>: Let $\mathfrak{C}$ and $\mathfrak{D}$ be equivalent categories of modules under the inverse equivalences $\mathfrak{F}\colon \mathfrak{C} \to \mathfrak{D}$ and $\mathfrak{G}\colon \mathfrak{D} \to \mathfrak{C}$. Then for any objects L and L' in $\mathfrak{C}$, the homomorphism from $\mathrm{Hom}_{\mathfrak{C}}(L,L')$ to $\mathrm{Hom}_{\mathfrak{D}}(\mathfrak{F}(L), \mathfrak{F}(L'))$ given by $g \to \mathfrak{F}(g)$ is one-one and onto.

<u>Proof</u>: Suppose f, g are elements of $\mathrm{Hom}_{\mathfrak{C}}(L,L')$ with $\mathfrak{F}(f) = \mathfrak{F}(g)$ in $\mathrm{Hom}_{\mathfrak{D}}(\mathfrak{F}(L), \mathfrak{F}(L'))$. Then $\mathfrak{G}(\mathfrak{F}(f)) = \mathfrak{G}(\mathfrak{F}(g)$ in $\mathrm{Hom}_{\mathfrak{C}}(\mathfrak{G}(\mathfrak{F}(L)), \mathfrak{G}(\mathfrak{F}(L')))$ which by the natural equivalence of $\mathfrak{G}\circ\mathfrak{F}$ with the identity functor implies that $f = g$. By a symmetric argument we see that $\mathfrak{G}\colon \mathrm{Hom}_{\mathfrak{D}}(\mathfrak{F}(L), \mathfrak{F}(L')) \to \mathrm{Hom}_{\mathfrak{C}}(\mathfrak{G}(\mathfrak{F}(L)), \mathfrak{G}(\mathfrak{F}(L')))$ is one-one.

Now suppose g is any element of $\mathrm{Hom}_{\mathfrak{D}}(\mathfrak{F}(L), \mathfrak{F}(L'))$. We then obtain the square

$$\begin{array}{ccc} \mathfrak{G}(\mathfrak{F}(L)) & \xrightarrow{\ \mathfrak{G}(g)\ } & \mathfrak{G}(\mathfrak{F}(L')) \\ {\scriptstyle \varphi_L}\downarrow & & \downarrow{\scriptstyle \varphi_{L'}} \\ L & \xrightarrow{\ f\ } & L' \end{array}$$

where φ_L and $\varphi_{L'}$ arise from the natural equivalence of $\mathfrak{G}\circ\mathfrak{F}$ with the identity and where $f = \varphi_{L'}\,\mathfrak{G}(g)\varphi_L^{-1}$. On the other hand, we also have the square

$$\begin{array}{ccc} \mathfrak{G}(\mathfrak{F}(L)) & \xrightarrow{\ \mathfrak{G}(\mathfrak{F}(f))\ } & \mathfrak{G}(\mathfrak{F}(L')) \\ {\scriptstyle \varphi_L}\downarrow & & \downarrow{\scriptstyle \varphi_{L'}} \\ L & \xrightarrow{\ f\ } & L' \end{array}$$

from which we deduce that $\mathfrak{G}(g) = \mathfrak{G}(\mathfrak{F}(f))$. Since $\mathfrak{G}$ is one-one, it follows that $g = \mathfrak{F}(f)$, so $\mathfrak{F}$ is onto. The proposition is proved.

For any ring R and any R-module M, set $M^* = \mathrm{Hom}_R(M,R)$ and $S = \mathrm{Hom}_R(M,M)$. Since R is an R-R-bimodule, M^* is a right R-module under the operation $(f\cdot r)(m) = f(m)r$. Moreover, M is a left S-module with $s\cdot m = s(m)$. Under this operation M is a left R- left S bimodule, whence M^* becomes a right S-module under the operation $(f\cdot s)(m) = f(s(m))$. One checks that M^* is in fact a right R- right S bimodule. It follows that we can form $M^* \otimes_D M$ and $M^* \otimes_C M$.

Moreover $M^* \otimes_R M$ is a left S- right S bimodule by virtue of M being a left R- left S bimodule and M^* being a right R- right S bimodule. Similarly $M^* \otimes_S M$ is a left R- right R bimodule.

Let θ_R denote the map from $M^* \otimes_R M$ to $S = \mathrm{Hom}_R(M,M)$ given by $[\theta_R(\sum_i f_i \otimes m_i)](m) = \sum_i f_i(m)m_i$. The reader should check that θ_R is both a left and a right S-module homomorphism.

Let θ_S denote the map from $M^* \otimes_S M$ to R given by $\theta_S(\sum_i f_i \otimes m_i) = \sum_i f_i(m_i)$. θ_S is a right and left R-module homomorphism whose image is the trace ideal $\mathfrak{T}_R(M)$.

Lemma 3.2: a.) θ_R is onto if and only if M is finitely generated and projective. If θ_R is onto, it is one-one.

b.) θ_S is onto if and only if M is a generator.

If θ_S is onto, it is one-one.

Proof: a.) Suppose θ_R is onto. Then there exists $\sum_{i=1}^{n} f_i \otimes m_i$ in $M^* \otimes_R M$ such that $\theta_R(\sum_{i=1}^{n} f_i \otimes m_i)$ is the identity map from M to M, that is, $\sum_{i=1}^{n} f_i(m)m_i = m$ for every $m \in M$. Thus the set $f_1,\ldots,f_n$, $m_1,\ldots, m_n$ forms a finite dual basis for M, so by the Dual Basis Lemma (1.3), M is finitely generated and projective. Conversely, supposing there exists a dual basis $f_1,\ldots, f_n \in M^*$, $m_1,\ldots, m_n \in M$, we see that $\theta_R(\sum_{i=1}^{n} (f_i g) \otimes m_i) = g$ for any g in $S = \mathrm{Hom}_R(M,M)$. Thus θ_R is onto.

Now given that θ_R is onto, we know M possesses a dual basis $f_1,\ldots, f_n \in M^*$, $m_1,\ldots, m_n \in M$. Suppose $\alpha = \sum_j h_j \otimes n_j$ is an element of $M^* \otimes_R M$ with $\theta_R(\alpha) = 0$, that is, $\sum_j h_j(m)n_j = 0$ for every m in M. Then

$$\alpha = \sum_j h_j \otimes n_j = \sum_j h_j \otimes (\sum_i f_i(n_j)m_i) = \sum_{i,j} h_j \otimes f_i(n_j)m_i = \sum_{i,j} (h_j \cdot f_i(n_j)) \otimes m_i$$

$= \sum_i (\sum_j h_j \cdot f_i(n_j)) \otimes m_i = 0$, since for any i, $[\sum_j h_j \cdot f_i(n_j)](m)$

$= \sum_j h_j(m) f_i(n_j) = f_i(\sum_j h_j(m)n_j) = f_i(0) = 0$ for every m in M.

This proves a.).

b.) Because the image of θ_S equals $\mathfrak{T}_R(M)$, the trace ideal of M, it is clear that θ_S is onto if and only if $\mathfrak{T}_R(M) = R$, that is, if and only if M is an R-generator.

Suppose θ_S is onto and $\sum_j h_j \otimes n_j$ is an element of $M^* \otimes_S M$ with $\theta_S(\sum_j h_j \otimes n_j) = 0$, that is, $\sum_j h_j(n_j) = 0$. Since θ_S is onto, there exist $f_1, \ldots, f_n$ in M^*, $m_1, \ldots, m_n$ in M with $\sum_i f_i(m_i) = 1$. Hence

$$\sum_j h_j \otimes n_j = \sum_j h_j \otimes (\sum_i f_i(m_i))n_j = \sum_{i,j} h_j \otimes \theta_R(f_i \otimes n_j)(m_i)$$

$$= \sum_i(\sum_j h_j \cdot \theta_R(f_i \otimes n_j)) \otimes m_i = 0,$$

since for every i and every m in M, $[\sum_j h_j \cdot \theta_R(f_i \otimes n_j)](m) = \sum_j h_j(f_i(m)n_j) = f_i(m)\sum_j h_j(n_j) = 0$. This completes the proof of the lemma.

For any left R-module M, we have seen that M is a left R- left S bimodule and $M^* = \mathrm{Hom}_R(M,R)$ is a right R- right S bimodule, where $S = \mathrm{Hom}_R(M,M)$. Therefore for any right R-module L, $L \otimes_R M$ has the structure of a left S-module, while for any left S-module N, $M^* \otimes_S N$ has the structure of a right R-module. Thus we have functors $(\ ) \otimes_R M: \mathfrak{M}_R \to {}_S\mathfrak{M}$ and $M^* \otimes_S (\ ): {}_S\mathfrak{M} \to \mathfrak{M}_R$. The following is the crucial proposition.

<u>Proposition 3.3</u>: Let R be any ring and let M be a left R-progenerator. Set $S = \mathrm{Hom}_R(M,M)$ and $M^* = \mathrm{Hom}_R(M,R)$. Then

$$(\ ) \otimes_R M: \mathfrak{M}_R \to {}_S\mathfrak{M}$$

and

$$M^* \otimes_S (\ ): {}_S\mathfrak{M} \to \mathfrak{M}_R$$

are inverse equivalences, establishing $\mathfrak{M}_R \sim {}_S\mathfrak{M}$.

<u>Proof</u>: We must show that each of the two ways of composing the above functors gives a functor naturally equivalent to the identity functor.

Let L be any object of $\mathfrak{M}_R$. Then, by the basic properties of the tensor product and Lemma 3.2b, we have $M^* \otimes_S (L \otimes_R M) \cong M^* \otimes_S (M \otimes_{R^\circ} L) \cong (M^* \otimes_S M) \otimes_{R^\circ} L \cong R \otimes_{R^\circ} L \cong L \otimes_R R \cong L$ where the composite isomorphism is given by $f \otimes (\ell \otimes m) \to \ell \cdot \theta_S (f \otimes m) = \ell \cdot f(m)$. This isomorphism allows one to verify that $(\ ) \otimes_R M$ followed by $M^* \otimes_S (\ )$ is naturally equivalent to the identity functor on $\mathfrak{M}_R$.

Similarly Lemma 3.2a yields that for any left S-module N, $(M^* \otimes_S N) \otimes_R M \cong (N \otimes_{S^\circ} M^*) \otimes_R M \cong N \otimes_{S^\circ} (M^* \otimes_R M) \cong N \otimes_{S^\circ} S \cong S \otimes_S N \cong N$ under the map $(f \otimes n) \otimes m \to \theta_R (f \otimes m) \cdot n$. Again this gives us that $M^* \otimes_S (\ )$ followed by $(\ ) \otimes_R M$ is naturally equivalent to the identity on ${}_S\mathfrak{M}$. Thus $(\ ) \otimes_R M$ and $M^* \otimes_S (\ )$ are inverse equivalences and the proposition is proved.

<u>Corollary 3.4</u>: In the setting of the proposition, we have

a.) $R \cong \mathrm{Hom}_S(M,M)$ (as rings) under the mapping which associates to an element r of R the endomorphism of M induced by scalar multiplication by r.

b.) $M^* \cong \mathrm{Hom}_S(M,S)$ (as right S-modules) under the mapping which associates to an element f of M^* the homomorphism $\theta_R(f \otimes (\ ))$ from M to S.

c.) $M \cong \mathrm{Hom}_R(M^*,R) = M^{**}$ (as left R-modules) under the mapping which associates to an element m in M the element in M^{**} which takes $f \in M^*$ to $f(m)$.

d.) $S^\circ \cong \mathrm{Hom}_R(M^*,M^*)$ (as rings) under the mapping which associates an element s of S° to the homomorphism from M^* to M^* given by $f \to fs$.

e.) M is an S-progenerator.

f.) M^* is an R-progenerator.

g.) M^* is an S-progenerator.

<u>Proof</u>: Under the equivalence of $\mathfrak{M}_R$ and ${}_S\mathfrak{M}$, the right R-module R corresponds to the left S-module $R \otimes_R M \cong M$ and the right R-module M^* corresponds to the left S-module $M^* \otimes_R M \cong S$. Therefore we can apply

Proposition 3.1 to obtain

a.) $R \cong \mathrm{Hom}_R(R,R) \cong \mathrm{Hom}_S(M,M)$

b.) $M^* \cong \mathrm{Hom}_R(R,M^*) \cong \mathrm{Hom}_S(M,S)$

c.) $M \cong \mathrm{Hom}_S(S,M) \cong \mathrm{Hom}_R(M^*,R)$

and d.) $S^\circ \cong \mathrm{Hom}_S(S,S) \cong \mathrm{Hom}_R(M^*,M^*)$.

The reader is urged to check that the composite isomorphisms are in each case the maps described in the statement of the corollary.

Now because M is an R-progenerator, we have $M^* \otimes_S M \cong R$ under θ_S and $M^* \otimes_R M \cong S$ under θ_R. By a.) and b.) above, this gives rise to isomorphisms

$$\mathrm{Hom}_S(M,S) \otimes_S M \cong \mathrm{Hom}_S(M,M)$$

and

$$\mathrm{Hom}_S(M,S) \otimes_{\mathrm{Hom}_S(M,M)} M \cong S.$$

Moreover the diligent reader will verify that the isomorphisms so obtained are exactly those of Lemma 3.2 (with R and S interchanged), so we conclude by Lemma 3.2 that M is an S-progenerator.

Again using $M^* \otimes_S M \cong R$ and $M^* \otimes_R M \cong S$ and this time substituting c.) and d.), we obtain

(′) $$R \cong M^* \otimes_S M \cong M^* \otimes_S \mathrm{Hom}_R(M^*,R)$$
$$\cong \mathrm{Hom}_R(M^*,R) \otimes_{S^\circ} M^* \cong \mathrm{Hom}_R(M^*,R) \otimes_{\mathrm{Hom}_R(M^*,M^*)} M^*$$

and

(″) $$\mathrm{Hom}_R(M^*,R) \otimes_{R^\circ} M^* \cong M^* \otimes_R \mathrm{Hom}_R(M^*,R) \cong M^* \otimes_R M \cong$$
$$S \cong \mathrm{Hom}_R(M^*,M^*) \cong \mathrm{Hom}_{R^\circ}(M^*,M^*)$$

where the last isomorphism in the second string is set identity and M^* is considered as a left R°-module since it is a right R-module. Again checking that the composite isomorphisms are those of Lemma 3.2 (with M^* in place of M), we see that M^* is an R-generator by (′) and a finitely generated and projective left R°-module by (″). Clearly this implies that M^* is a right R-progenerator.

Finally since M is an S-progenerator by e.), we can apply f.) to the S-module M to conclude that $\mathrm{Hom}_S(M,S)$ is an S-progenerator. But by b.) $\mathrm{Hom}_S(M,S)$ is isomorphic as a right S-module to M^*, so M^* must be an S-progenerator.

Corollary 3.5: Let the setting be as in the proposition. For any two-sided ideal $\mathfrak{a}$ of R, $M^*\otimes_R(\mathfrak{a}\otimes_R M)$ is naturally isomorphic to the two-sided ideal of S consisting of all elements of the form $\sum_i \theta_R(f_i\otimes \alpha_i m_i)$, $f_i\in M^*$, $\alpha_i\in\mathfrak{a}$, $m_i\in M$. Similarly for any two-sided ideal $\mathfrak{b}$ of S, $M^*\otimes_S(\mathfrak{b}\otimes_S M)$ is naturally isomorphic to the two-sided ideal of R consisting of all elements of the form $\sum_i \theta_S(f_i\otimes\beta_i(n_i)) = \sum_i f_i(\beta_i(n_i))$, $f_i\in M^*$, $\beta_i\in\mathfrak{b}$, $n_i\in M$. These correspondences are inverses of each other and establish a one-one, order preserving correspondence between the two-sided ideals of R and the two-sided ideals of S.

Proof: We know that M and M^* are both R-projective and so are flat. Thus the exact sequence $0\to\mathfrak{a}\to R$ yields $0\to M^*\otimes_R(\mathfrak{a}\otimes_R M)\to M^*\otimes_R(R\otimes_R M)\cong M^*\otimes_R M\cong S$. Considering $M^*\otimes_R(\mathfrak{a}\otimes_R M)$ therefore as contained in $M^*\otimes_R(R\otimes_R M)$, the isomorphism of $M^*\otimes_R(R\otimes_R M)$ with S takes $M^*\otimes_R(\mathfrak{a}\otimes_R M)$ onto the ideal of S made up of elements of the form $\sum_i \theta_R(f_i\otimes\alpha_i m_i)$.

Similarly since M and M^* are S-projective, $M^*\otimes_S(\mathfrak{b}\otimes_S M)$ can be considered inside $M^*\otimes_S(S\otimes_S M)$ which is isomorphic to R under a map which takes $M^*\otimes_S(\mathfrak{b}\otimes_S M)$ to the ideal of R made up of elements looking like $\sum_i f_i(\beta_i(n_i))$.

Next we observe that $\mathfrak{a}\to M^*\otimes_R(\mathfrak{a}\otimes_R M)\to M^*\otimes_S(M^*\otimes_R\mathfrak{a}\otimes_R M)\otimes_S M$ corresponds to the two-sided ideal $\mathfrak{a}'$ of R consisting of all elements of the form $\sum_i f_i(m_i)\alpha_i g_i(n_i)$ where f_i, g_i are from M^*, α_i is from $\mathfrak{a}$ and m_i, n_i are from M. Clearly $\mathfrak{a}'\subset\mathfrak{a}$. Moreover, since there exist $f_i\in M^*$ and $m_i\in M$ with $\sum_i f_i(m_i) = 1$, we have that any α in $\mathfrak{a}$ can be written as $\sum_i f_i(m_i)\alpha f_i(m_i)$ and so is in $\mathfrak{a}'$. Thus $\mathfrak{a}=\mathfrak{a}'$. A similar argument using a dual basis shows that the two-sided ideal of S corresponding to $M^* \otimes_R (M^* \otimes_S \mathfrak{b} \otimes_S M) \otimes_R M$ equals $\mathfrak{b}$. Thus the correspondences are mutual inverses and the corollary is proved.

We should perhaps point out that certain well known classical results follow immediately from Corollaries 3.4 and 3.5. We note two of the most obvious, namely, that the double dual of a finite dimensional vector space is naturally isomorphic to the vector space (3.4c) and that the n-by-n matrices over a field form a simple ring (3.5).

We conclude this section with a final corollary which holds in a general categorical setting. However we state and prove it in our special context.

Corollary 3.6: In the setting of the proposition, a right R-module L is an R-progenerator if and only if its corresponding left S-module $L \otimes_R M$ is an S-progenerator.

Proof: Suppose L is an R-progenerator. Then L is a direct summand of a free R-module of finite rank, that is, there exists an R-module L' and a positive integer n with $L \oplus L' \cong R^{(n)}$. Then as left S-modules we have $(L \oplus L')\otimes_R M \cong (L\otimes_R M) \oplus (L'\otimes_R M) \cong R^{(n)}\otimes_R M \cong M\oplus M\oplus\cdots\oplus M$ (n-times). But by Corollary 3.4 M is finitely generated and projective as an S-module, so $L \otimes_R M$ is a direct summand of the finitely generated, projective S-module $M\oplus\cdots\oplus M$ and so is itself finitely generated and projective over S.

Also L is a generator, so there exist $f_1, \ldots, f_n \in L^*$, $\ell_1, \ldots, \ell_n \in L$ with $\sum_i^n f_i(\ell_i) = 1$. M is an S-generator, so there exist $g_1, \ldots, g_k \in \mathrm{Hom}_S(M,S)$ and $m_1, \ldots, m_k \in M$ with $\sum_j^k g_j(m_j) = 1$. Define h_{ij} in $\mathrm{Hom}_S(L\otimes_R M, S)$ by $h_{ij}(\ell\otimes m) = g_j(f_i(\ell)m)$. Then $\sum_{i,j} h_{ij}(\ell_i \otimes m_j) = \sum_j g_j(\sum_i f_i(\ell_i)m_j) = \sum_j g_j(m_j) = 1$, whence $L \otimes_R M$ is an S-generator.

By symmetry we have that L is a progenerator over R if $L \otimes_R M$ is one over S.

§ 4. Local Rings; Rings and Modules of Quotients; Localization.

The ultimate aim of this section is to prove that the concept of the rank of a free module can be extended to finitely generated, projective modules over commutative rings containing no idempotents but 0 and 1. To do this we must introduce the important technique of localization.

We call a (not necessarily noetherian) commutative ring *local* if it contains a unique maximal ideal. Clearly a field is a local ring. We will soon show that there is associated to every prime ideal $\mathfrak{p}$ of a commutative ring R a local ring, denoted $R_{\mathfrak{p}}$, which is a ring of quotients of R in much the same way that the rational numbers are a ring of quotients of the rational integers.

One of the most useful properties of a local ring is the fact that projective modules are free. We prove this for finitely generated modules.

Proposition 4.1: Let R be a local ring with maximal ideal $\mathfrak{m}$ and let M be a finitely generated, projective R-module. Suppose $m_1 + \mathfrak{m}M, \ldots, m_n + \mathfrak{m}M$ is a free basis for $M/\mathfrak{m}M$ over $R/\mathfrak{m}$. Then $m_1, \ldots, m_n$ is a free basis for M over R.

Proof: Let $\varphi\colon R^{(n)} \to M$ be defined by $\varphi((\alpha_1, \ldots, \alpha_n)) = \sum_{i=1}^{n} \alpha_i m_i$. Since $m_1 + \mathfrak{m}M, \ldots, m_n + \mathfrak{m}M$ generate $M/\mathfrak{m}M$, we have $Rm_1 + \cdots + Rm_n + \mathfrak{m}M = M$. Therefore by applying Corollary 1.8 to the module $M/(Rm_1 + \cdots + Rm_n)$, we have $Rm_1 + \cdots + Rm_n = M$, so φ is onto.

Now suppose $(\alpha_1, \ldots, \alpha_n)$ is in the kernel of φ. Then $\sum_i \alpha_i m_i = 0$, so $\sum_i \alpha_i (m_i + \mathfrak{m}M) = 0$ in $M/\mathfrak{m}M$. Thus each α_i is in $\mathfrak{m}$ since the $m_i + \mathfrak{m}M$ are free over $R/\mathfrak{m}$, so $\ker(\varphi) \subset \mathfrak{m}R^{(n)}$. Since M is projective, the exact sequence

$$0 \to \ker(\varphi) \to R^{(n)} \xrightarrow{\varphi} M \to 0$$

splits, so there exists a submodule L of $R^{(n)}$ with $\ker(\varphi) \oplus L = R^{(n)}$. Consequently $\ker(\varphi) \subset \mathfrak{m}R^{(n)} \cap \ker(\varphi) = (\mathfrak{m}\ker(\varphi) \oplus \mathfrak{m}L) \cap \ker(\varphi) = \mathfrak{m}\ker(\varphi)$. But $\ker(\varphi)$, being a direct summand of $R^{(n)}$, is finitely generated and so equals (0) by Corollary 1.8. Thus φ is an isomorphism, proving the proposition.

We next present a general method of constructing modules from a given module patterned after the construction of the rationals from the integers.

We call a subset W of a commutative ring R <u>multiplicatively closed</u> if

a.) $1 \in W$

and

b.) w, w' in W implies ww' in W.

In our considerations, multiplicatively closed sets will consist either of all the non-negative powers of an element of R or of all the elements of R not contained in some proper prime ideal.

Suppose M is any R-module and W is any multiplicatively closed subset of R. We introduce a relation $\approx$ into the set $M \times W$ by defining $(m_1, w_1) \approx (m_2, w_2)$ if and only if there exists $w \in W$ with $w(w_1m_2 - w_2m_1) = 0$. One sees that $\approx$ is an equivalence relation. We denote the equivalence class containing (m, w) by $\frac{m}{w}$ and the collection of all equivalence classes by M_W. Define addition and R-scalar multiplication in M_W by $\frac{m_1}{w_1} + \frac{m_2}{w_2} = \frac{w_1m_2 + w_2m_1}{w_1w_2}$ and $r(\frac{m}{w}) = \frac{rm}{w}$. We encourage the reader to check that these definitions are independent of the choice of representatives of the equivalence classes and that M_W thereby forms an R-module. M_W is called a <u>module of quotients</u>.

If, in addition, M is an R-algebra, M_W can be checked to be an R-algebra under the multiplication $\frac{m_1}{w_1} \cdot \frac{m_2}{w_2} = \frac{m_1m_2}{w_1w_2}$. In particular, R_W is called a <u>ring of quotients</u> of R. Specializing further, if we take W to be $R-\mathfrak{p}$, where $\mathfrak{p}$ is a prime ideal of R, we denote R_W by $R_{\mathfrak{p}}$ (not $R_{R-\mathfrak{p}}$) and observe that the ideal $\mathfrak{p}R_{\mathfrak{p}} = \{ \frac{r}{s} \mid r \in \mathfrak{p},\ s \in R-\mathfrak{p}\}$ of $R_{\mathfrak{p}}$ has the property that every element of $R_{\mathfrak{p}}$ outside of $\mathfrak{p}R_{\mathfrak{p}}$ is a unit of $R_{\mathfrak{p}}$. Therefore every proper ideal of $R_{\mathfrak{p}}$ is contained in $\mathfrak{p}R_{\mathfrak{p}}$, whence $\mathfrak{p}R_{\mathfrak{p}}$ is the unique maximal ideal of $R_{\mathfrak{p}}$. Thus $R_{\mathfrak{p}}$ is a local ring, called the <u>localization of R at the prime $\mathfrak{p}$</u>.

We now prove some fundamental properties of rings and modules of quotients.

<u>Lemma 4.2</u>: Let W be any multiplicatively closed subset of R. Then for any R-module M, M_W is an R_W-module and the mapping from M_W to $M \otimes_R R_W$ given by $\frac{m}{w} \to m \otimes \frac{1}{w}$ is an R_W-module isomorphism.

<u>Proof</u>: One checks that the operation defined by $\frac{r}{w} \cdot \frac{m}{w'} = \frac{rm}{ww'}$ is an R_W-scalar multiplication of M_W which makes M_W an R_W-module.

It is not hard to show that the map $\frac{m}{w} \to m \otimes \frac{1}{w}$ is indeed a well-defined R_W-homomorphism. Moreover, the map from $M \otimes_R R_W$ back to M_W given by $m \otimes \frac{r}{w} \to \frac{rm}{w}$ is well-defined and obviously the inverse of the preceding map. Thus $R \otimes_R R_W \cong M_W$.

In what follows we will sometimes casually and without notice shift between M_W and $M \otimes_R R_W$.

<u>Lemma 4.3</u>: R_W is a flat R-module.

<u>Proof</u>: In view of Lemma 4.2, it is enough to show that if $f\colon M \to N$ is an R-module monomorphism from the R-module M to the R-module N, then the map $f_W\colon M_W \to N_W$ given by $f_W(\frac{m}{w}) = \frac{f(m)}{w}$ is an R_W-monomorphism. That f_W is an R_W-homomorphism is clear. Suppose $\frac{f(m)}{w} = 0$. Then there exists some w' in W with $w'f(m) = f(w'm) = 0$. But f is a monomorphism, so $w'm = 0$, implying $\frac{m}{w} = 0$ in R_W.

We can now prove the useful

<u>Proposition 4.4:</u> Let M be an R-module such that $M_{\mathfrak{m}} = 0$ for every maximal ideal $\mathfrak{m}$ of R. Then M = 0.

Proof: Let m be any element of M. It sufficies to show m = 0. For each maximal ideal $\mathfrak{m}$ of R, $\frac{m}{1} = 0$ in $M_{\mathfrak{m}}$, so there exists an element $\alpha_{\mathfrak{m}}$ in R-$\mathfrak{m}$ with $\alpha_{\mathfrak{m}} m = 0$. Thus the annihilator ideal of m in R is contained in no maximal ideal and so equals R. Hence M = 0.

We would like to generalize the concept of the rank of a free module to projective modules. Suppose M is a finitely generated, projective R-module, where R is any commutative ring. For any prime ideal $\mathfrak{p}$ of R, we can form the $R_{\mathfrak{p}}$-module $M \otimes_R R_{\mathfrak{p}} \cong M_{\mathfrak{p}}$. By earlier results (2.2 and 4.1), we know that $M \otimes_R R_{\mathfrak{p}}$ is $R_{\mathfrak{p}}$-free, so there exists a unique non-negative integer $n_{\mathfrak{p}}$ with $M \otimes_R R_{\mathfrak{p}} \cong R_{\mathfrak{p}}^{(n_{\mathfrak{p}})}$. We call $n_{\mathfrak{p}}$ the $\mathfrak{p}$-rank of M and denote it $\mathrm{rank}_{\mathfrak{p}}(M)$. For a non-negative integer n, we call n the rank of M and write $\mathrm{rank}_R(M) = n$ if and only if $\mathrm{rank}_{\mathfrak{p}}(M) = n$ for every prime ideal $\mathfrak{p}$ of R; that is, we define the rank of M only if all the $\mathfrak{p}$-ranks are equal.

Proposition 4.5: Let R be a commutative ring and M a finitely generated, projective R-module such that $\mathrm{rank}_R(M)$ is defined. Then for any commutative R-algebra S, $\mathrm{rank}_S(M \otimes_R S)$ is defined and equals $\mathrm{rank}_R(M)$.

Proof: Let $\mathfrak{P}$ be any prime ideal of S. It is clear that $\mathfrak{p} = \{x \in R \mid x \cdot 1 \in \mathfrak{P}\}$ is a prime ideal of R. Moreover one checks that $S_{\mathfrak{P}}$ is naturally an $R_{\mathfrak{p}}$-algebra since $(R-\mathfrak{p}) \cdot 1 \subset S-\mathfrak{P}$. (see 4.8) Therefore $(M \otimes_R S) \otimes_S S_{\mathfrak{P}} \cong M \otimes_R (S \otimes_S S_{\mathfrak{P}}) \cong M \otimes_R S_{\mathfrak{P}} \cong M \otimes_R (R_{\mathfrak{p}} \otimes_{R_{\mathfrak{p}}} S_{\mathfrak{P}}) \cong (M \otimes_R R_{\mathfrak{p}}) \otimes_{R_{\mathfrak{p}}} S_{\mathfrak{P}}$ $\cong R_{\mathfrak{p}}^{(n)} \otimes_{R_{\mathfrak{p}}} S_{\mathfrak{P}} \cong S_{\mathfrak{P}}^{(n)}$ where $n = \mathrm{rank}_R(M)$. Since $\mathfrak{P}$ is arbitrary, this proves the proposition.

The remainder of this section is devoted to proving that when R contains no idempotents other than 0 and 1, the rank of any finitely generated, projective R-module is defined.

For any commutative ring R, we call the collection of all prime ideals of R which do not equal R the spectrum of R and denote it

Spec(R). We introduce a topology, the Zariski topology, into Spec(R) as follows: for any subset L of R, let $h(L) = \{\mathfrak{p} \in \mathrm{Spec}(R) \mid \mathfrak{p} \supset L\}$. If we designate the subsets h(L) of Spec(R) as closed sets, we have

Lemma 4.6: The subsets h(L) form a topology for Spec(R).

Proof: It is trivial to verify that $h(R) = \emptyset$, $h(0) = \mathrm{Spec}(R)$, and $\bigcap_i h(L_i) = h(\bigcup_i L_i)$. Moreover, since we are dealing with prime ideals, we can check that

$h(L_1) \cup h(L_2) = h(\{\ell_1\ell_2 \mid \ell_1 \in L_1, \ell_2 \in L_2\})$ for any subsets L_1, L_2 of R. This establishes the lemma.

To prove the next proposition, we remind the reader of some useful facts about commutative rings.

a.) If $\mathfrak{a}$ and $\mathfrak{b}$ are comaximal ideals of a commutative ring R, that is, if $\mathfrak{a} + \mathfrak{b} = R$, then $\mathfrak{a}^m$ and $\mathfrak{b}^n$ are comaximal for any positive integers m and n and $\mathfrak{a}^m \cap \mathfrak{b}^n = \mathfrak{a}^m\mathfrak{b}^n$.

b.) (Krull) The intersection of all the prime ideals of a commutative ring is exactly the set of nilpotent elements in the ring. (See Zariski-Samuel, [L] pp. 151-152.)

Proposition 4.7: The topological space Spec(R) is connected if and only if 0 and 1 are the only idempotents in R.

Proof: Recall that a space is disconnected if and only if there exist two non-empty disjoint closed subsets whose union is the entire space.

Suppose e is an idempotent in R equal neither to 1 nor 0. Then h(e) and h(1-e) are non-empty (since $Re \neq R$ and $R(1-e) \neq R$) and closed. Since $(1-e)e = 0$, every prime ideal contains either e or 1-e, so $h(e) \cup h(1-e) = \mathrm{Spec}(R)$. Finally an element of Spec(R) in both h(e) and h(1-e) would contain $1 = (1-e) + e$, which can't be, so $h(e) \cap h(1-e) = \emptyset$, proving Spec(R) disconnected.

Now suppose conversely that Spec(R) is disconnected; that is, there exist $h(L_1)$ and $h(L_2)$ disjoint, non-empty closed subsets of Spec(R) with $h(L_1) \cup h(L_2) = \mathrm{Spec}(R)$. Set $\mathfrak{a}_1 = \bigcap_{\mathfrak{p} \in h(L_1)} \mathfrak{p}$ and $\mathfrak{a}_2 = \bigcap_{\mathfrak{p} \in h(L_2)} \mathfrak{p}$. It is clear that $h(\mathfrak{a}_1) = h(L_1) \neq \emptyset$ and $h(\mathfrak{a}_2) = h(L_2) \neq \emptyset$, so $\mathfrak{a}_1 \neq R$ and $\mathfrak{a}_2 \neq R$. Also, since $h(\mathfrak{a}_1) \cap h(\mathfrak{a}_2) = h(L_1) \cap h(L_2) = \emptyset$, there can be no primes containing both $\mathfrak{a}_1$ and $\mathfrak{a}_2$, whence $\mathfrak{a}_1 + \mathfrak{a}_2 = R$, that is, $\mathfrak{a}_1$ and $\mathfrak{a}_2$ are comaximal. Furthermore, $h(\mathfrak{a}_1) \cup h(\mathfrak{a}_2) = \mathrm{Spec}(R)$ implies $\mathfrak{a}_1 \cap \mathfrak{a}_2$ is the intersection of all the prime ideals of R and so consists of nilpotent elements by Krull's result. If we write $1 = \alpha_1 + \alpha_2$ with $\alpha_i \in \mathfrak{a}_i$, then $R = R\alpha_1 + R\alpha_2$, so $R\alpha_1$ and $R\alpha_2$ are comaximal with $\alpha_1\alpha_2 \in \mathfrak{a}_1 \cap \mathfrak{a}_2$, so $(\alpha_1\alpha_2)^m = 0$ for some positive integer m. Thus $R\alpha_1^m$ and $R\alpha_2^m$ are comaximal with $R\alpha_1^m \cap R\alpha_2^m = 0$, giving $R = R\alpha_1^m \oplus R\alpha_2^m$. But any non-trivial decomposition of R into a direct sum of ideals yields idempotents different from 0 and 1, so the proposition is proved.

We return to consideration of rings of quotients.

<u>Lemma 4.8</u>: Let W and W' be multiplicatively closed subsets of R with $W \subset W'$. Then the map from R_W to $R_{W'}$ given by $\frac{r}{w} \to \frac{r}{w}$ is a ring homomorphism which induces in $R_{W'}$ the structure of an R_W-algebra. Moreover, $R_{W'} \cong (R_W)_{W'R_W}$ where $W'R_W = \{ \frac{w'}{w} \in R_W \mid w' \in W' \}$ is a multiplicatively closed subset of R_W.

<u>Proof</u>: The fact that $W \subset W'$ gives that the map $\frac{r}{w} \to \frac{r}{w}$ is well-defined (where of course the first $\frac{r}{w}$ is considered an element in R_W and the second $\frac{r}{w}$ an element of $R_{W'}$). To show the isomorphism between $R_{W'}$ and $(R_W)_{W'R_W}$, one can define a map which takes $\frac{r}{w'}$ of $R_{W'}$ to $\dfrac{\frac{r}{1}}{\frac{w'}{1}}$ in $(R_W)_{W'R_W}$ and show it is well-defined and an isomorphism. We omit the messy and uninstructive details.

Lemma 4.9: If M is a finitely generated R-module and W is a multiplicatively closed subset of R, then $M_W = 0$ if and only if there exists an element $w \in W$ with $wM = 0$.

Proof: Clearly if $wM = 0$, then any element $\frac{m}{w}$, of R_W equals zero.

Conversely, suppose $M_W = 0$ and $M = Rm_1 + \cdots + Rm_n$. Since $\frac{m_i}{1} = 0$ in M_W, there exists $w_i \in W$ with $w_i m_i = 0$. Then $w = \prod_{i=1}^{n} w_i$ has the property that $wm_i = 0$ for every i, so $wM = 0$.

We have seen that for a finitely generated, projective R-module M, $M_{\mathfrak{p}}$ is free over $R_{\mathfrak{p}}$ for any prime ideal $\mathfrak{p}$. Our next proposition shows that we can take a multiplicatively closed subset considerably smaller than $R-\mathfrak{p}$ and obtain the same result.

For any element α in R, we let $R_{(\alpha)}$ denote the ring of quotients of R determined by the multiplicatively closed subset of R consisting of the non-negative powers of α.

Proposition 4.10: Let M be a finitely generated, projective R-module and $\mathfrak{p}$ any prime ideal of R. Then there exists $\alpha \in R-\mathfrak{p}$ with $M \otimes_R R_{(\alpha)} \cong M_{(\alpha)}$ free as an $R_{(\alpha)}$-module.

Proof: We know that $M_{\mathfrak{p}}$ is free of finite rank. Let $\frac{m_1}{\alpha_1}, \ldots, \frac{m_n}{\alpha_n}$ form a basis for $M_{\mathfrak{p}}$ over $R_{\mathfrak{p}}$. Since $\frac{1}{\alpha_1}, \ldots, \frac{1}{\alpha_n}$ are units in $R_{\mathfrak{p}}$, it follows that $\frac{m_1}{1}, \ldots, \frac{m_n}{1}$ also form a basis for $M_{\mathfrak{p}}$ over $R_{\mathfrak{p}}$.

Define an R-homomorphism φ: $R^{(n)} \to M$ by $\varphi((\rho_1, \ldots, \rho_n)) = \sum_{i=1}^{n} \rho_i m_i$. This gives rise to an exact sequence

$$(*) \qquad 0 \to \ker(\varphi) \to R^{(n)} \overset{\varphi}{\to} M \to \operatorname{coker}(\varphi) \to 0.$$

Since $R_{\mathfrak{p}}$ is R-flat (4.3), we obtain the exact sequence

$$0 \to \ker(\varphi)_{\mathfrak{p}} \to [R^{(n)}]_{\mathfrak{p}} \to M_{\mathfrak{p}} \to \operatorname{coker}(\varphi)_{\mathfrak{p}} \to 0 .$$

Because $\{\frac{m_1}{1}, \ldots, \frac{m_n}{1}\}$ is a basis, $\ker(\varphi)_{\mathfrak{p}} = \operatorname{coker}(\varphi)_{\mathfrak{p}} = 0$. Because $\operatorname{coker}(\varphi)$ is finitely generated, Lemma 4.9 insures the existence of $\beta \in R-\mathfrak{p}$ with $\beta[\text{coker}(\varphi)] = 0$, so $[\text{coker}(\varphi)]_{(\beta)} = 0$. Thus we obtain from (*) the exact sequence

$$0 \to [\ker(\varphi)]_{(\beta)} \to [R^{(n)}]_{(\beta)} \to M_{(\beta)} \to 0 .$$

Next we see that $[\ker(\varphi)]_{(\beta)}$ must be finitely generated over $R_{(\beta)}$ since the preceding sequence splits, $M_{(\beta)}$ being $R_{(\beta)}$-projective by Corollary 2.2. But by Lemma 4.8, $R_{\mathfrak{p}}$ is an $R_{(\beta)}$-algebra and we have $[\ker(\varphi)]_{(\beta)} \otimes_{R_{(\beta)}} R_{\mathfrak{p}} \cong (\ker(\varphi) \otimes_R R_{(\beta)} \otimes_{R_{(\beta)}} R_{\mathfrak{p}} \cong \ker(\varphi) \otimes_R (R_{(\beta)} \otimes_{R_{(\beta)}} R_{\mathfrak{p}}$ $\cong \ker(\varphi)_{\mathfrak{p}} = 0$, so again by Lemma 4.9, there exists $\frac{\mu}{\beta^k}$ in $R_{(\beta)} - \mathfrak{p}R_{(\beta)}$ with $\frac{\mu}{\beta^k}[\ker(\varphi)_{(\beta)}] = 0$. Since $\frac{1}{\beta^k}$ is a unit in $R_{(\beta)}$, this is equivalent to $\mu[\ker(\varphi)_{(\beta)}] = 0$. We see that $[R_{(\beta)}]_{(\frac{\mu}{1})}$ is isomorphic to $R_{(\mu\beta)}$, so we have

$$0 = [\ker(\varphi)_{(\beta)}]_{(\frac{\mu}{1})} \cong (\ker(\varphi) \otimes_R R_{(\beta)}) \otimes_{R_{(\beta)}} [R_{(\beta)}]_{(\frac{\mu}{1})} \cong \ker(\varphi) \otimes_R R_{(\mu\beta)}$$

$\cong \ker(\varphi)_{(\mu\beta)}$. Since $\beta[\text{coker}(\varphi)] = 0$, we also have $\text{coker}(\varphi)_{(\mu\beta)} = 0$. Thus we obtain from (*) the exact sequence

$$0 \to [R^{(n)}]_{(\mu\beta)} \to M_{(\mu\beta)} \to 0 ,$$

proving that for $\alpha = \mu\beta$, $M_{(\alpha)}$ is $R_{(\alpha)}$-free.

<u>Corollary 4.11</u>: Let M be a finitely generated, projective R-module. The map from Spec(R) to $\{0, 1, 2, 3, \ldots\}$ given by $\mathfrak{p} \to \text{rank}_{\mathfrak{p}}(M)$ is continuous when the non-negative integers are given the discrete topology.

<u>Proof</u>: Suppose α is any element of R such that $M_{(\alpha)}$ is free of rank m over $R_{(\alpha)}$. Then, for any prime ideal $\mathfrak{p}$ of R not containing α, we have that $R_{\mathfrak{p}}$ is an $R_{(\alpha)}$-algebra and $R_{\mathfrak{p}} \otimes_R M \cong R_{\mathfrak{p}} \otimes_{R_{(\alpha)}} (R_{(\alpha)} \otimes_R M)$ $\cong R_{\mathfrak{p}} \otimes_{R_{(\alpha)}} [R_{(\alpha)}]^{(m)} \cong [R_{\mathfrak{p}}]^{(m)}$, so $\text{rank}_{\mathfrak{p}}(M) = m$.

Now let $\mathfrak{p}$ be any prime and suppose $\text{rank}_{\mathfrak{p}}(M) = n$. By the preceding proposition, there exists $\alpha \in R-\mathfrak{p}$ with $M_{(\alpha)}$ free of finite rank over $R_{(\alpha)}$. By the preceding paragraph, this rank must equal n. Consider

$\text{Spec}(R)-h(\alpha)$. Since $h(\alpha)$ is closed, $\text{Spec}(R)-h(\alpha)$ is an open subset of $\text{Spec}(R)$ containing $\mathfrak{p}$. Again by the preceding paragraph, if $\mathfrak{q} \in \text{Spec}(R)-h(\alpha)$, $\text{rank}_{\mathfrak{q}}(M) = n = \text{rank}_{\mathfrak{p}}(M)$, so the preimage of n is open in $\text{Spec}(R)$. Therefore the map is continuous.

We can now deduce the following theorem as an immediate corollary of this result and Proposition 4.7, recalling that the continuous image of a connected space is connected.

Theorem 4.12: Let R be a commutative ring with no idempotents but 0 and 1. Then, for any finitely generated, projective R-module M, $\text{rank}_R(M)$ is defined; that is, there exists a non-negative integer m such that $\text{rank}_{\mathfrak{p}}(M) = m$ for every prime ideal $\mathfrak{p}$ of R.

§ 5. The Projective Class Group

In this section a covariant functor from the category of commutative rings (and ring homomorphisms) to the category of abelian groups (and group homomorphisms) is described. To each commutative ring R will be associated an abelian group P(R) called the class group of R. If R is a Dedekind domain a concrete realization of P(R) is provided.

Lemma 5.1: Let E be a finitely generated projective faithful module over the commutative ring R. Then the following are equivalent:

1. $\text{Rank}_R(E) = 1$
2. $\text{Rank}_R(\text{Hom}_R(E,R)) = 1$
3. $\text{Hom}_R(E,E) \cong R$
4. $\text{Hom}_R(E,R) \otimes_R E \cong R$.

Proof: ($1 \Longleftrightarrow 2$) By Corollary 2.5 we have for each prime ideal $\mathfrak{p}$ of R that $R_{\mathfrak{p}} \otimes \text{Hom}_R(E,R) = \text{Hom}_{R_{\mathfrak{p}}}(E_{\mathfrak{p}},R_{\mathfrak{p}})$. Any finitely generated projective $R_{\mathfrak{p}}$-module is free (4.1) so if $E_{\mathfrak{p}} \cong R_{\mathfrak{p}}^{(n)}$ we know $\text{Hom}_{R_{\mathfrak{p}}}(E_{\mathfrak{p}},R_{\mathfrak{p}}) \cong \text{Hom}_{R_{\mathfrak{p}}}(R_{\mathfrak{p}}^{(n)},R_{\mathfrak{p}}) = R_{\mathfrak{p}}^{(n)}$. Therefore $\text{Rank}_R(E) = 1$ if and only if

$\text{Rank}_R(\text{Hom}_R(E,R)) = 1$.

(1→3) Define ϕ: $R \to \text{Hom}_R(E,E)$ by $\phi(r)[e] = r \cdot e$. One can check that for each $r \in R$, $\phi(r) \in \text{Hom}_R(E,E)$. Also ϕ is a ring homomorphism from R to $\text{Hom}_R(E,E)$. Now ϕ is a monomorphism since E is a faithful R-module. For each prime ideal $\mathfrak{p}$ of R there is induced the commuting diagram

$$\begin{array}{ccc} R_{\mathfrak{p}} \otimes_R R & \xrightarrow{1 \otimes \phi} & R_{\mathfrak{p}} \otimes_R \text{Hom}_R(E,E) \\ \downarrow & & \downarrow \\ R_{\mathfrak{p}} & \xrightarrow{\phi_{\mathfrak{p}}} & \text{Hom}_{R_{\mathfrak{p}}}(R_{\mathfrak{p}}, R_{\mathfrak{p}}) \end{array}$$

The vertical arrows are the natural isomorphisms and $\phi_{\mathfrak{p}}$ is an isomorphism so $1 \otimes \phi$ is an isomorphism. We have shown that $R_{\mathfrak{p}} \otimes_R$ image ϕ $= R_{\mathfrak{p}} \otimes_R \text{Hom}_R(E,E)$ for all prime ideals $\mathfrak{p}$ of R so by (4.4), image ϕ $= \text{Hom}_R(E,E)$ and ϕ is an isomorphism.

(3 → 4) This is Lemma 3.2a .

(4 → 1) For each prime ideal $\mathfrak{p}$ of R we note that $(R_{\mathfrak{p}} \otimes_R (\text{Hom}_R(E,R) \otimes_R E)) = R_{\mathfrak{p}}$. But $R_{\mathfrak{p}} \otimes_R (\text{Hom}_R(E,R) \otimes_R E) \cong R_{\mathfrak{p}} \otimes_R \text{Hom}_R(E,R) \otimes_{R_{\mathfrak{p}}} (R_{\mathfrak{p}} \otimes_R E) \cong \text{Hom}_{R_{\mathfrak{p}}}(E_{\mathfrak{p}}, R_{\mathfrak{p}}) \otimes_{R_{\mathfrak{p}}} (E_{\mathfrak{p}})$. Thus for all prime ideals $\mathfrak{p}$ of R, $1 = \text{Rank}_{R_{\mathfrak{p}}}(E_{\mathfrak{p}})$ and therefore $\text{Rank}_R(E) = 1$.

Given a commutative ring R let P(R) be the set of isomorphism classes of finitely generated projective faithful R-modules satisfying any of the four conditions of Lemma 5.1. To see that P(R) is indeed a well defined set check the argument in Section 5 of Chapter II. Multiply two equivalence classes $|E_1|$ and $|E_2|$ by $|E_1| \cdot |E_2| = |E_1 \otimes_R E_2|$. The fundamental properties of the tensor product and rank insure that this is a well defined associative commutative product with identity $|R|$. Lemma 5.1 guarantees for each $|E| \in P(R)$ the inverse $|\text{Hom}_R(E,R)| \in P(R)$ so P(R) is an abelian group, called the <u>projective class group</u>.

Let f: R → S be a ring homomorphism from R to the commutative ring

ring S, then we can think of S as an R-algebra via f and define a homomorphism from P(R) to P(S) denoted P(f) by $P(f)(|E|) = |S \otimes_R E|$. The verification that P(f) is a well defined homomorphism is routine.

It is straightforward now to verify that we have defined a covariant functor from the category of commutative rings (and ring homomorphisms) to the category of abelian groups (and group homomorphisms).

There is a special case where the class group can be more concretely realized. Let R be a domain with quotient field K. Let I(R) be the set of all finitely generated R-submodules of K (i.e. all fractional ideals) and let Pr(R) be the elements of I(R) which are generated by a single element (i.e. principal fractional ideals). The domain R is called a Dedekind domain in case R is noetherian and every element of I(R) is projective over R. (For other equivalent definitions see Vol. 1 of Zariski-Samuel's Commutative Algebra [L]). Let R be a Dedekind domain. If $E_1 \in I(R)$ then $K \otimes_R E_1 \cong K$ by $a \otimes e \to ae$ so $\mathrm{Rank}_R(E_1) = 1$. Moreover if $E_2 \in I(R)$ then $E_1 \otimes_R E_2 \cong E_1E_2$ under $a \otimes b \to ab$. Thus I(R) is a commutative semi-group with identity R and there is a homomorphism $E \to |E|$ from I(R) to P(R) with kernel Pr(R). If E is a finitely generated projective R-module and $\mathrm{Rank}_R(E) = 1$ then $K \otimes_R E \cong K$ so E can be identified with a fractional ideal and every element in P(R) is the image of some element in I(R). Consequently, since P(R) is a group, if $E_1 \in I(R)$ there is an element $E_2 \in I(R)$ with $E_1E_2 \in Pr(R)$. If $E_1E_2 = \{rk \mid r \in R\}$ then $\{r \frac{1}{k} \mid r \in R\} = I$ is an element of Pr(R) and $E_1(E_2I) = R$. This proves I(R) is a group and Pr(R) is a subgroup. Collecting all the facts above we have shown

<u>Theorem 5.2</u>: Let R be a Dedekind domain, let I(R) be the group of fractional ideals of R and Pr(R) the subgroup of principal fractional ideals. Then if P(R) is the class group of R we have

$$P(R) \cong I(R)/Pr(R) .$$

The results of this section make it clear that if R is a local ring or a principal ideal domain then P(R) is trivial. In general the computation of P(R) is a difficult task.

Exercises

1.) Let A be a finitely generated R-algebra.

a.) Show $\text{rad}(R)\cdot A \subset \text{rad}(A)$, where rad denotes the Jacobson radical.

b.) Show that there exists a positive integer n such that $[\text{rad}(A)]^n \subset \bigcap_{\mathfrak{m}} (\mathfrak{m}A)$ where $\mathfrak{m}$ runs over the maximal ideals of R.

c.) If A is R-projective, $\text{rad}(R)\cdot A = \bigcap(\mathfrak{m}A)$.

2.) <u>Transitivity of Rank</u>

a.) Let A be a finitely generated, commutative algebra over the field R. Let M be a finitely generated, projective A-module. Set $n = \text{rank}_A M$ (suppose it is defined) and $m = \text{rank}_R(A)$. Show $\text{rank}_R(M) = mn$.

b.) Let R be a commutative ring now and M any finitely generated, projective R-module such that $\text{rank}_R(M)$ is defined. Show that for any maximal ideal $\mathfrak{m}$ of R, $\text{rank}_R(M) = \text{rank}_{R/\mathfrak{m}}(M/\mathfrak{m}M)$.

c.) Prove that if A is a commutative, finitely generated, projective R-algebra and M is a finitely generated, projective A-module such that $\text{rank}_A(M)$ and $\text{rank}_R(A)$ are defined, then $\text{rank}_R(M)$ is defined and equals $\text{rank}_R(A)\cdot\text{rank}_A(M)$.

3.) Let R be a ring and M an R-module. Show that the following are equivalent:

a.) M is an R-generator.

b.) R is a direct summand of a finite direct sum of

c.) For any R-modules N and N' and g, g' in $Hom_R(N,N')$, $g \neq g'$ implies there exists f in $Hom_R(M,N)$ with $gf \neq g'f$.

4.) Use exercise 3 to show that generator modules are preserved under any category equivalence; i.e., if $\mathfrak{F}$ is a category equivalence from one category of modules to another, then a module L is a generator in the first category if and only if $\mathfrak{F}(L)$ is a generator in the second category.

5.) a.) In any category, a map $f:M \rightarrow N$ is called a monomorphism if whenever L is an object of the category and g,h are maps from L to M, then $fg = fh$ implies $g = h$.

Show that in any category consisting of all the modules over some ring and all the homomorphisms between modules, a map is a monomorphism if and only if it is one-one.

b.) In any category, a map $f:M \rightarrow N$ is called a proper monomorphism if f is a monomorphism but there is no map $g:N \rightarrow M$ such that $fg = i_N$ and $gf = i_M$. Two monomorphisms $f:M \rightarrow N$ and $g:L \rightarrow N$ are ordered $f \leq g$ if and only if there exists a map $h:M \rightarrow L$ with $f = gh$.

Show that in any category consisting of all the modules over some ring, the following statements about a fixed module M are equivalent:

i.) M is finitely generated;

ii.) the union of every linearly ordered chain of proper submodules of M is a proper submodule;

iii.) every linearly ordered chain of proper monomorphisms into M possesses an upper bound in the collection of all proper monomorphisms into M.

6.) Use exercises 4 and 5 to prove that if $\mathfrak{M}_R$ and ${}_S\mathfrak{M}$ are equivalent under a category equivalence $\mathfrak{F}$, then a module M in $\mathfrak{M}_R$ is an

R-progenerator if and only if $\mathfrak{F}(M)$ is an S-progenerator.

7.) <u>Converse to Proposition 3.3</u>

Suppose $\mathfrak{M}_R$ and ${}_S\mathfrak{M}$ are equivalent under $\mathfrak{F}$: $\mathfrak{M}_R \to {}_S\mathfrak{M}$ and $\mathfrak{G}$: ${}_S\mathfrak{M} \to \mathfrak{M}_R$, where R and S are rings. By exercise 6 $\mathfrak{F}(R)$ is an S-progenerator and $\mathfrak{G}(S)$ is an R-progenerator. Show that $\mathfrak{F}(\ ) = (\ )\otimes_R\mathfrak{F}(R)$ and $\mathfrak{G}(\ ) = \mathfrak{G}(S)\otimes_S(\ )$ with $S \cong \mathrm{Hom}_R(\mathfrak{F}(R), \mathfrak{F}(R))$ and $\mathfrak{G}(S) \cong \mathrm{Hom}_R(\mathfrak{F}(R), R)$ in the following steps, giving a converse to 3.3.

a.) Show that $R \cong \mathrm{Hom}_R(R,R) \overset{\mathfrak{F}}{\to} \mathrm{Hom}_S(\mathfrak{F}(R), \mathfrak{F}(R))$ endows $\mathfrak{F}(R)$ with the structure of a left R-left S bimodule. Prove a similar result for $\mathfrak{G}(S)$.

b.) Let A and B be rings. Let P be any finitely generated, projective left B-module. Let $Q \in {}_S\mathfrak{M}_R$ and $T \in \mathfrak{M}_R$. Show that the map Φ from $\mathrm{Hom}_A(Q,T) \otimes_B P$ to $\mathrm{Hom}_A(\mathrm{Hom}_B(P,Q),T)$ given by $\Phi(f\otimes p)(g) = f(g(p))$ is an isomorphism.

c.) Show $\mathrm{Hom}_S(\mathfrak{F}(R), \mathfrak{F}(R)) \cong R$.

d.) Use c.) and Proposition 3.3 to get $S \cong \mathrm{Hom}_R(\mathfrak{F}(R), \mathfrak{F}(R))$ and $\mathfrak{G}(S) \cong \mathrm{Hom}_R(\mathfrak{F}(R), R)$.

e.) Show that $\mathfrak{F}(L) = L\otimes_R\mathfrak{F}(R)$ by writing $L\otimes_R\mathfrak{F}(R)$ as $\mathrm{Hom}_R(R,L) \otimes_R \mathfrak{F}(R)$ and using b.), d.) and Proposition 3.1.

8.) Let R be a commutative ring and M a finitely generated, free R-module. Show that any two bases of M consist of the same number of elements. Show that this need not be true for free modules of finite rank over non-commutative rings.

9.) Show that in the context of Corollary 3.4, $R \cong \mathrm{Hom}_S(M^*,M^*)$ and $M \cong \mathrm{Hom}_S(M^*,S)$. Determine the maps.

10.) Let A be a finitely generated, projective R-algebra, R a commutative ring. Show that if elements x and y of A have the property $xy = 1$, then $yx = 1$. [Consider the map $f: A \to A$ given by $f(a) = ay$. Show it is onto and therefore one-one].

11.) a.) Let R be a commutative ring and $\varphi: M \to N$ an R-module homomorphism. Then φ is one-one (resp. onto) if and only if $\varphi_{\mathfrak{m}}: M_{\mathfrak{m}} \to N_{\mathfrak{m}}$ is one-one (resp. onto) for every maximal ideal $\mathfrak{m}$ of R.

b.) Let A be an R-algebra and $M \in \mathfrak{M}_A$, $N \in {}_A\mathfrak{M}$. Show that for every maximal ideal $\mathfrak{m}$ of R, $(M_{\mathfrak{m}} \otimes_{A_{\mathfrak{m}}} N_{\mathfrak{m}}) \cong (M \otimes_A N)_{\mathfrak{m}}$.

c.) If $f: R \to S$ is a ring homomorphism of not necessarily commutative rings and M is a flat R-module, show that $M \otimes_R S$ is a flat S-module.

d.) Flatness is a local property. Let A be an R-algebra. Then the following statements concerning the left R-module M are equivalent:

i.) M is flat;

ii.) $M_{\mathfrak{m}}$ is $A_{\mathfrak{m}}$-flat for every maximal ideal $\mathfrak{m}$ of R.

12.) Let M be a finitely generated, projective R-module of rank 1. Show that M is faithful.

13.) Show that if S is a faithfully flat R-algebra and M is an R-module, then M is R-flat if and only if $M \otimes_R S$ is S-flat.

CHAPTER II

In this chapter we present the fundamental theory of non-commutative separable algebras, with the attendant theory of the Brauer group of a commutative ring. The primary reference for this material is M. Auslander and O. Goldman's "The Brauer Group of a Commutative Ring" [7] which appeared in 1960. This theory successfully generalizes the classical theory of central simple algebras over fields and the Brauer group of a field. The reader is urged to consult the monograph by Artin, Nesbitt and Thrall [B] for a direct exposition of the theory for fields. Throughout this chapter A will denote a not necessarily commutative R-algebra. The unadorned symbol "⊗" will always mean tensoring with respect to R.

§ 1. Definitions, Examples, and Basic Properties

This section begins with the fundamental definition of separability. The fact that an algebra over a field is separable in the present sense if and only if it is finite dimensional and separable in the classical sense (see Section 2) is the crucial connection with the classical theory.

For any R-algebra A we can form the so-called enveloping algebra $A \otimes A^{\circ}$ of A, where A° denotes the R-algebra opposite to A.[1] For convenience we will write A^e for the enveloping algebra $A \otimes A^{\circ}$ of A.

The algebra A has a structure as a left A^e-module induced by $(a \otimes a')\cdot b = aba'$. Also there is a map μ from the algebra A^e onto A given by $\mu(\sum_i a_i \otimes a_i') = \sum_i a_i a_i'$. We see that μ is a left A^e-module homomorphism, which in case A is commutative is a ring homomorphism. Therefore, setting $J = \text{kernel}\ (\mu)$, we obtain the exact sequence

(1 - A° is the ring having the same underlying additive group as A with multiplication $*$ defined by $a * b = ba$ where ba denotes the product of b with a in the ring A. However, we will suppress the "$*$" since the context will make it clear whether we are operating in A or A°.)

$$0 \to J \to A^e \overset{\mu}{\to} A \to 0$$

of left A^e-modules. The reader can verify that J is the left ideal of A^e generated by all elements of the form $a \otimes 1 - 1 \otimes a$.

Proposition 1.1: The following conditions on an R-algebra A are equivalent:

i) A is projective as a left A^e-module under the μ-structure.

ii) $0 \to J \to A^e \overset{\mu}{\to} A \to 0$ splits as a sequence of left A^e-modules.

iii) A^e contains an element e such that $\mu(e) = 1$ and $Je = 0$.

Proof: The equivalence of i) and ii) is clear. If we assume the sequence splits, there exists a left A^e-homomorphism ψ from A to A^e such that $\mu\psi$ is the identity on A. Denoting $\psi(1)$ by e, we see that $\mu(e) = 1$ and $Je = 0$. Conversely, if A^e contains an element e such that $\varphi(e) = 1$ and $Je = 0$, the map $\psi\colon A \to A^e$ defined by $\psi(a) = (a \otimes 1)e = (1 \otimes a)e$ is an A^e-homomorphism because $\psi(a \otimes b \cdot c) = \psi(acb) = (acb \otimes 1)e = (ac \otimes 1)(b \otimes 1)e = (ac \otimes 1)(1 \otimes b)e = (ac \otimes b)e = (a \otimes b)(c \otimes 1)e = a \otimes b \cdot \psi(c)$. Moreover, $\mu\psi$ is the identity on A, so ψ is a splitting map. This proves the equivalence of ii) and iii).

Definition of Separability: An R-algebra A is called separable if it satisfies the equivalent conditions of Proposition 1.1.

We should remark that the element e described in condition iii) of the proposition is necessarily an idempotent, since $e^2 - e = (e - 1 \otimes 1)e \in J\cdot e = 0$. We will refer to e as a separability idempotent for A. Notice that a separability idempotent for an algebra lies not in the algebra itself but in its enveloping algebra.

We now give a list of three examples of separable algebras.

Example I: Let R be a field. An R-algebra A is called classically separable if the Jacobson radical of $A \otimes_R K$ is zero for every field extension K of R. Then A is separable if and only if A is classically separable and the dimension of A as a vector space over R is finite. (See § 2 for proof.)

<u>Example II</u>: Let $M_n(R)$ denote the ring of all $n \times n$ matrices over R. Let e_{ij} denote the matrix having 1 in the (i,j)-th spot and 0 elsewhere. Set $e = \sum_i e_{ij} \otimes e_{ji}$ for any fixed j between 1 and n. Then $\mu(e) = \sum_i e_{ij}e_{ji} = \sum_i e_{ii} = 1$ and for any k and ℓ,

$$(e_{k\ell} \otimes 1 - 1 \otimes e_{k\ell})e = \sum_i (e_{k\ell}e_{ij} \otimes e_{ji} - e_{ij} \otimes e_{ji}e_{k\ell})$$

$$= e_{kj} \otimes e_{j\ell} - e_{kj} \otimes e_{j\ell} = 0 .$$

Since the $e_{k\ell}$ generate $M_n(R)$ as an R-module, this shows that $Je = 0$, so e is a separability idempotent for $M_n(R)$, that is $M_n(R)$ is a separable R-algebra.

<u>Example III:</u> Let G be a finite group whose order n is a unit in R. Then the group algebra R(G) is a separable R-algebra, for if we set $e = \frac{1}{n} \sum_\sigma \sigma \otimes \sigma^{-1}$ in $[R(G)]^e$, we have $\mu(e) = 1$ and for any $\tau \in G$,

$$(\tau \otimes 1)e = \frac{1}{n} \sum_\sigma \tau\sigma \otimes \sigma^{-1} = \frac{1}{n} \sum_\omega \omega \otimes \omega^{-1} \tau = (1 \otimes \tau)e .$$

We now formulate a useful module-theoretic condition equivalent to separability. For any R-algebra A, it is apparent that a left A-module M inherits an R-module structure when we define $r \cdot m = (r \cdot 1) \cdot m$ and similarly $m \cdot r = m \cdot (r \cdot 1)$ for right A-modules. By a <u>two-sided A/R-module</u> we mean an abelian group M along with commuting left and right A-module structures whose induced R-module structures coincide. More explicitly, a two-sided A/R-module M is a left A-module which is also a right A-module such that

i) $(a \cdot m) \cdot a' = a \cdot (m \cdot a')$, for all m in M, a, a′ in A; and

ii) $(r \cdot 1) \cdot m = m \cdot (r \cdot 1)$, for all m in M, r in R.

If M is a left A^e-module, M can be regarded as a two-sided A/R-module by defining

$$a \cdot m = (a \otimes 1)m \text{ and } m \cdot a = (1 \otimes a)m$$

and conversely any two-sided A/R-module can be considered a left A^e-module. Therefore the concepts of a left A^e-module and that of a two-sided A/R-module are equivalent and we will use them interchangeably.

For any two-sided A/R-module M, let $\underline{M^A}$ denote the set $\{m \in M \mid a \cdot m = m \cdot a, \ \forall\, a \in A\}$. M^A is an R-submodule of M, and one can check that $(\ \ )^A$ is a covariant functor from the category of two-sided A/R-modules to the category of R-modules.

<u>Lemma 1.2</u>: For any two-sided A/R-module M, $\mathrm{Hom}_{A^e}(A,M) \simeq M^A$ as R-modules under the correspondence $f \to f(1)$ where $f \in \mathrm{Hom}_{A^e}(A,M)$. For any other two-sided A/R-module N and any g in $\mathrm{Hom}_{A^e}(M,N)$, the diagram

$$\begin{array}{ccc} \mathrm{Hom}_{A^e}(A,M) & \xrightarrow{g \circ (\ )} & \mathrm{Hom}_{A^e}(A,N) \\ \downarrow & & \downarrow \\ M^A & \xrightarrow{g|} & N^A \end{array}$$

commutes.

<u>Proof</u>: If f is in $\mathrm{Hom}_{A^e}(A,M)$, $a \cdot f(1) = a \otimes 1 \cdot f(1) = f(a \otimes 1 \cdot 1)$ $= f(a) = f(1 \otimes a \cdot 1) = 1 \otimes a \cdot f(1) = f(1) \cdot a$, so $f(1) \in M^A$. Conversely, for any $m \in M^A$, the function f defined by $f(a) = a \cdot m$ is an element of $\mathrm{Hom}_{A^e}(A,M)$. Therefore the mapping $f \to f(1)$ is one-one and onto M^A. The rest follows immediately.

In categorical terms Lemma 1.2 establishes that the left exact covariant functors $\mathrm{Hom}_{A^e}(A,\ \ )$ and $(\ \ )^A$ are naturally equivalent.

<u>Corollary 1.3</u>: $\mathrm{Hom}_{A^e}(A,A) \simeq Z(A)$, the center of A, under the correspondence $f \to f(1)$.

<u>Proof:</u> Set $A = M$ in the lemma and note that $A^A = Z(A)$.

Corollary 1.4: Let $0:J$ denote the right annihilator of J in A^e. Then $\mathrm{Hom}_{A^e}(A, A^e) \simeq 0:J$ and, if A is R-separable, $\mu(0:J) = Z(A)$.

Proof: Setting $M = A^e$ in the lemma, we have $\mathrm{Hom}_{A^e}(A, A^e) \simeq (A^e)^A = \{x \in A^e \mid (a \otimes 1 - 1 \otimes a)x = 0, \ \forall\, a \in A\} = 0:J$. Also, $A^e \xrightarrow{\mu} A \to 0$ exact implies $\mathrm{Hom}_{A^e}(A, A^e) \xrightarrow{\mu\circ(\ )} \mathrm{Hom}_{A^e}(A, A) \to 0$ exact if A is A^e-projective. [$\mathrm{Hom}_{A^e}(A, \ )$ is an exact functor if and only if A is A^e-projective.] Then applying the lemma yields $\mu(0:J) = Z(A)$.

Corollary 1.5: An R-algebra A is separable if and only if $(\ )^A$ is a right exact functor; that is, if and only if $M \xrightarrow{f} N \to 0$ exact implies $M^A \xrightarrow{f|} N^A \to 0$ exact, for every pair of two-sided A/R-modules M and N and every A^e-epimorphism f.

Proof: $\mathrm{Hom}_{A^e}(A, \ )$ is right exact if and only if A is A^e-projective, that is, if and only if A is R-separable. But Lemma 1.2 yields that $(\ )^A$ is right exact if and only if $\mathrm{Hom}_{A^e}(A, \ )$ is.

We now need to investigate the behavior of separable algebras under certain formal operations, for example the taking of a tensor product or the formation of a factor algebra. The next five propositions set out the crucial properties and in each case we use the separability criterion of Corollary 1.5 as the tool of proof.

Proposition 1.6: Let S_1 and S_2 be commutative R-algebras. Let A_1 be a separable S_1-algebra and A_2 a separable S_2-algebra. Then $A_1 \otimes A_2$ is a separable $S_1 \otimes S_2$-algebra under the operation induced by $(s_1 \otimes s_2)\cdot(a_1 \otimes a_2) = s_1 \cdot a_1 \otimes s_2 \cdot a_2$ (provided $A_1 \otimes A_2 \neq 0$ and $S_1 \otimes S_2 \neq 0$).

Proof: Let $M \xrightarrow{f} N \to 0$ be an exact sequence of two-sided $A_1 \otimes A_2/S_1 \otimes S_2$-modules. Then the natural map $A_1 \to A_1 \otimes A_2$ given by $a_1 \to a_1 \otimes 1$ endows M and N with a structure as two-sided A_1/S_1-modules.

Therefore $M^{A_1} \xrightarrow{f|} N^{A_1} \to 0$ is exact by Corollary 1.5 since A_1 is S_1-separable. Since the natural image of A_2 in $A_1 \otimes A_2$ commutes with the image of A_1, M^{A_1} and N^{A_1} are two-sided A_2/S_2-modules under the operation induced from $1 \otimes A_2$, so again applying Corollary 1.5, we have

$$(M^{A_1})^{A_2} \xrightarrow{f|} (N^{A_1})^{A_2} \to 0$$

is exact by the S_2-separability of A_2. But it is clear that $(M^{A_1})^{A_2} = M^{A_1 \otimes A_2}$ and $(N^{A_1})^{A_2} = N^{A_1 \otimes A_2}$, so

$$M^{A_1 \otimes A_2} \to N^{A_1 \otimes A_2} \to 0$$

is exact, implying that $A_1 \otimes A_2$ is $S_1 \otimes S_2$-separable.

<u>Corollary 1.7</u>: Let A be a separable R-algebra and S any commutative R-algebra. Then $A \otimes S$ is separable as an S-algebra.

<u>Proof</u>: Set $A = A_1$, $R = S_1$, $S = S_2 = A_2$ in Proposition 1.6.

We can also prove a converse to Proposition 1.6.

<u>Proposition 1.8</u>: Let S_1 and S_2 be commutative R-algebras. Let A_1 be any S_1-algebra and A_2 any S_2-algebra such that $A_1 \otimes A_2$ is separable as an $S_1 \otimes S_2$-algebra. Then, if A_2 is faithful as an R-module and contains R as an R-direct summand, A_1 is separable over S_1.

<u>Proof</u>: If M is any two-sided A_1/S_1-module, $M \otimes A_2$ is a two-sided $A_1 \otimes A_2/S_1 \otimes S_2$-module. Since R is an R-direct summand of A_2, $M \otimes A_2 \cong (M \otimes L) \oplus (M \otimes R)$ (as two-sided A_1/S_1-modules) for some R-submodule L of A_2, so M can be identified with a direct summand of $M \otimes A_2$. Moreover, since the projection of $M \otimes A_2$ onto M is a two-sided A_1/S_1-homomorphism, it is easily seen that under this map, $(M \otimes A_2)^{A_1 \otimes A_2}$ is projected onto M^{A_1}.

Now consider the commutative diagram

$$\begin{array}{ccccc} M\otimes A_2 & \xrightarrow{f\otimes 1} & N\otimes A_2 & \longrightarrow & 0 \\ \downarrow & & \downarrow & & \\ M & \xrightarrow{f} & N & \longrightarrow & 0 \end{array}$$

where the vertical maps are the projections and the rows are exact. This gives rise to the commutative diagram

$$\begin{array}{ccccc} (M\otimes A_2)^{A_1\otimes A_2} & \xrightarrow{f\otimes 1} & (N\otimes A_2)^{A_1\otimes A_2} & \longrightarrow & 0 \\ \downarrow & & \downarrow & & \\ M^{A_1} & \longrightarrow & N^{A_1} & & \end{array}$$

where the vertical maps are still projections and the top row is exact since $A_1 \otimes A_2$ is $S_1 \otimes S_2$-separable. Now it follows obviously that $M^{A_1} \to N^{A_1} \to 0$ is exact, so A_1 is S_1-separable.

<u>Corollary 1.9</u>: If A_1 and A_2 are R-algebras with $A_1 \otimes A_2$ separable over R and A_2 contains R as an R-direct summand, then A_1 is R-separable.

<u>Proof</u>: Take $S_1 = S_2 = R$ in the proposition.

<u>Corollary 1.10</u>: If A is an R-algebra and S is any commutative R-algebra containing R as an R-direct summand, then A is R-separable if $A \otimes S$ is S-separable. Moreover, if $i(1 \otimes S)$ is the center of $A \otimes S$, $R\cdot 1$ is the center of A.

<u>Proof</u>: To obtain the first conclusion, set $A_1 = A$, $S_1 = R$ and $A_2 = S_2 = S$ in the proposition. To see the second, we have by the proof of the proposition that $1 \otimes S = Z(A \otimes S) = (A \otimes S)^{A\otimes S}$ is projected onto $A^A = Z(A)$. But $1 \otimes S$ is projected onto $1 \otimes R \simeq R\cdot 1$.

Before stating the next proposition, we remark that if $\mathfrak{a}$ is any ideal of R such that $\mathfrak{a}\cdot 1 = 0$, then A is naturally an $R/\mathfrak{a}$-algebra. Moreover one sees that $A \otimes_R A^o = A \otimes_{R/\mathfrak{a}} A^o$, so A is R-separable if and only if A is $R/\mathfrak{a}$-separable. We will frequently make this shift of coefficient ring, sometimes without further comment.

Proposition 1.11: Let A be a separable R-algebra and let $\mathfrak{A}$ be a two-sided ideal of A. Then $A/\mathfrak{A}$ is a separable R-algebra (and hence, by the preceding remark, $A/\mathfrak{A}$ is a separable $R\cdot 1/R\cdot 1\cap\mathfrak{A}$-algebra.) Furthermore, $Z(A/\mathfrak{A}) = \frac{Z(A)+\mathfrak{A}}{\mathfrak{A}}$.

Proof: Every two-sided $(A/\mathfrak{A})/R$-module M is naturally a two-sided A/R-module under the operation $a\cdot m = (a + \mathfrak{A})\cdot m$ and $m\cdot a = m\cdot(a + \mathfrak{A})$. Under this arrangement, $M^A = M^{A/\mathfrak{A}}$, so $A/\mathfrak{A}$ is separable by a simple application of Corollary 1.5.

Now $A \to A/\mathfrak{A} \to 0$ is an exact sequence of two-sided A/R-modules, so separability of A implies

$$A^A \to (A/\mathfrak{A})^A \to 0$$

is exact. But $A^A = Z(A)$ and $(A/\mathfrak{A})^A = Z(A/\mathfrak{A})$, so $Z(A/\mathfrak{A})$ is the image of Z(A) under the natural map, that is, $Z(A/\mathfrak{A}) = \frac{Z(A)+\mathfrak{A}}{\mathfrak{A}}$.

Proposition 1.12: (Transitivity of Separability) Let S be a commutative, separable R-algebra and let A be a separable S-algebra. Then A is naturally an R-algebra and is R-separable. If, on the other hand, A is given to be a separable R-algebra and S is any R-subalgebra of the center of A, then A is separable over S.

Proof: Clearly the composition of the structure homomorphism of R into S with the structure homomorphism of S into the center of A endows A with the structure of an R-algebra.

Any two-sided A/R-module M is a fortiori a two-sided S/R-module. Moreover, for $x \in M^S$, $a \in A$ and $s \in S$,

$$s\cdot(a\cdot x) = a\cdot(s\cdot x) = a\cdot(x\cdot s) = (a\cdot x)\cdot s\ ,$$

so $ax \in M^S$. It follows that M^S is a two-sided A/S-module, with $(M^S)^A = M^A$. Therefore the separability of A over S and the separability of S over R gives by Corollary 1.5 that

$$M \xrightarrow{f} N \to 0 \qquad \text{exact}$$

implies $\quad (M^S)^A = M^A \xrightarrow{f|} (N^S)^A = N^A \to 0 \quad$ exact,

for any two-sided A/R-modules M and N. Hence A is R-separable.

We leave the verification of the final statement of the proposition to the reader.

Proposition 1.13: Let A_1 be an R_1-algebra and A_2 an R_2-algebra, where R_1 and R_2 are commutative rings. Then $A_1 \oplus A_2$ is a separable $R_1 \oplus R_2$-algebra if and only if both A_1 and A_2 are separable over R_1 and R_2 respectively.

Proof: Easy applications of Corollary 1.5 and Proposition 1.11.

§ 2. Separable Algebras over Fields

We devote this short section to proving the connection between the present and the classical definitions of separability in the case that the coefficient ring **R** is a field. We begin by proving a general result of Villamayor and Zelinsky [92].

Proposition 2.1: Let A be a separable R-algebra which is projective as an R-module. Then A is finitely generated as an R-module.

Proof: Let $\{f_i, a_i\}$ be a dual basis for A° over R with $a_i \in A^\circ$ and $f_i \in \mathrm{Hom}_R(A^\circ, R)$. (Clearly A° is projective as an R-module if A is.) Then $a = \sum_i f_i(a)\, a_i$ for every $a \in A^\circ$ where, for fixed a, $f_i(a) = 0$ for all but a finite number of subscripts i. If we identify $A \otimes_R R$ with A, $1_A \otimes f_i$ can be considered an element of $\mathrm{Hom}_A(A^e, A)$ and the set $\{1_A \otimes f_i, 1 \otimes a_i\}$ forms a dual basis for A^e as a projective left A-module, that is

$$u = \sum_i (1_A \otimes f_i)(u) \cdot (1 \otimes a_i) \qquad \text{for all } u \in A^e .$$

Applying the multiplication map μ and setting $u = (1 \otimes a)e$ where e is a separability idempotent for A over R, we obtain

$$a = \mu((1 \otimes a)e) = \sum_i [(1_A \otimes f_i)((1 \otimes a)e)]\cdot a_i . \qquad \text{(I)}$$

Since $(1_A \otimes f_i)((1 \otimes a)e) = (1_A \otimes f_i)((a \otimes 1)e) = (a \otimes 1)((1_A \otimes f_i)(e))$, the set of subscripts i for which $(1_A \otimes f_i)((1 \otimes a)e) \neq 0$ is contained in the finite set of subscripts for which $(1_A \otimes f_i)(e) \neq 0$ and this latter set is independent of a. Therefore the summation (I) may be taken over a fixed finite set. Writing $e = \sum_j x_j \otimes y_j$, we have from (I) that

$$a = \sum_{i,j} x_j \; f_i(y_j \; a)a_i = \sum_{i,j} f_i(y_j \; a)x_ja_i$$

so the finite set $\{x_ja_i\}$ generates A° (and therefore A) over R.

<u>Corollary 2.2</u>: Let A be a separable R-algebra where R is a field. Then the dimension of A as a vector space over R is finite.

We next prove that all separable algebras possess the important "lifting property" for projective modules.

<u>Proposition 2.3</u>: Let A be a separable R-algebra. Any left A-module M which is R-projective (under its induced R-module structure) is A-projective.

<u>Proof</u>: Let $0 \to L \to N \overset{\eta}{\to} M \to 0$ be any exact sequence of left A-modules. To show that M is A-projective it suffices to show this sequence A-splits. Supposing M to be R-projective, we know this sequence R-splits, that is, there exists an R-module homomorphism ψ from M to N with $\eta\psi = 1_M$.

Since both N and M are left A-modules, $\mathrm{Hom}_R(M,N)$ can be given the structure of a left A^e-module under the operation induced by

$$[(a \otimes a')\cdot f](m) = a\cdot f(a'\cdot m), \quad \forall a, a' \in A, \; f \in \mathrm{Hom}_R(M,N), \; m \in M.$$

We can now use the separability of A to modify ψ suitably in order to obtain a new map ψ' from M to N which still has the property that $\eta\psi' = 1_M$ but in addition is an A-homomorphism. Let e be a separability idempotent for A. Define $\psi' = e\cdot\psi$, that is, if $e = \sum_i x_i \otimes y_i$,

$$\psi'(m) = \sum_i x_i \ \psi(y_i m), \qquad \forall\ m \in M .$$

Since η is an A-homomorphism and $\mu(e) = 1$

$$\eta\ \psi'(m) = \eta\ (\sum_i x_i\ \psi(y_i m)) = \sum_i x_i\ \eta\cdot\psi(y_i m) = \sum_i x_i y_i\ m = m ,$$

$\forall\ m \in M$. Furthermore, since $Je = 0$, we have

$$(a \otimes 1-1 \otimes a)\psi' = (a \otimes 1-1 \otimes a)e\cdot\psi = 0 \quad , \forall\ a \in A ,$$

so $\qquad a\ \psi'(m) = a \otimes 1\cdot\psi'(m) = 1 \otimes a\cdot\psi'(m) = \psi'(a m),\ \forall a \in A,\ m \in M.$

Therefore ψ' is an A-homomorphism which completes the proof.

Corollary 2.4: If A is a separable R-algebra where R is a field, then A is classically separable as an R-algebra.

Proof: Recall that to show A classically separable we must prove that $A \otimes K$ has Jacobson radical equal to the zero ideal, for every field extension K of R. By Corollary 1.7 we know that $A \otimes K$ is a separable K-algebra. But every $A \otimes K$-module is K-projective (since every module over a field is free), from which it follows by the proposition that every $A \otimes K$-module is projective. Hence $A \otimes K$ has zero radical and the proof is complete. (Here we use the fact that a ring with the descending chain condition over which every module is projective has zero radical. See 25.8, page 166 of Curtis and Reiner [G].)

We now wish to prove the converse to Corollaries 2.2 and 2.4; that is, we show that if A is a finite dimensional, classically separable algebra over a field R, then A is separable.

Let S be an algebraic closure of R. If A is classically separable and of finite dimension over R, $A \otimes S$ is a ring with the descending chain condition having zero radical. Therefore by the Wedderburn-Artin theory $A \otimes S$ is a direct sum of a finite number of full matrix rings over division algebras, where each division algebra is finite dimensional over its center and each center contains (a copy of) S. But it is easy to see that there are no proper division algebras of finite dimension over an algebraically closed field, so $A \otimes S$ is a finite direct sum of full matrix rings over S. Now example II shows that each of these matrix rings is a separable S-algebra, so by Proposition 1.13 $A \otimes S$ is separable over $S \oplus \cdots \oplus S$ (n-times where n is the number of simple components of $A \otimes S$). However it is a minor exercise that $S \oplus \cdots \oplus S$ is separable over S, so $A \otimes S$ is separable over S by Proposition 1.12. It now follows from Corollary 1.10 that A is R-separable. Hence

Theorem 2.5: An algebra A over a field R is separable if and only if A is classically separable over R and the dimension of A over R is finite.

§ 3. Central Separable Algebras

We have seen (Proposition 1.12) that if A is a separable R-algebra, A is separable when considered as an algebra over its center. We now turn our attention to this type of algebra.

An R-algebra A is called central if A is faithful as an R-module and $R \cdot 1$ coincides with the center of A. When dealing with faithful algebras, we identify R with $R \cdot 1$ and therefore consider R as a subring of the center of A. We call A a central separable* R-algebra if A is both central and separable.

We should remark that a central separable algebra over a field is central simple because by Theorem 2.5 it is a direct sum of simple rings and it can have only one simple component since its center has only one. Conversely, any simple ring which is finite dimensional

(*-N. Bourbaki [E] calls central separable algebras Azumaya algebras due presumably to the remarkable work done by G. Azumaya in "On maximally central algebras" [8].)

over its center is central separable (Proposition 1.2 of Chapter V). Thus the theory of central separable algebras over a field R is the study of the central simple R-algebras of finite dimension.

For any R-algebra A we have seen that A is naturally a left A^e-module. This structure induces an R-algebra homomorphism φ from A^e to $\text{Hom}_R(A,A)$ by associating to any element α of A^e the element $\varphi(\alpha)$ of $\text{Hom}_R(A,A)$ which is scalar multiplication in A by α. More explicitly, if $\alpha = \sum_i a_i \otimes a_i'$, then $\varphi(\alpha)(a) = \alpha \cdot a = \sum_i a_i\, a\, a_i'$. We now set out to prove that in the case that A is a central separable R-algebra this algebra homomorphism φ is an isomorphism.

<u>Lemma 3.1</u>: Let A be a central separable R-algebra. Then R is an R-direct summand of A.

<u>Proof</u>: Let e be a separability idempotent for A (in A^e). Consider the homomorphism $\varphi(e)$ in $\text{Hom}_R(A,A)$ where φ is the map defined just above. Since φ is a ring homomorphism, $\varphi(e)$ is an idempotent. In other terms $\varphi(e)$ is a projection of A into A, from which it follows that the image of $\varphi(e)$ is an R-direct summand of A having the kernel of $\varphi(e)$ as its complementary direct summand. But remembering that $Je = 0$, we see that for any a, b $\in$ A,

$$a\cdot[\varphi(e)(b)] = (a \otimes 1)e\cdot b = (1 \otimes a)e\cdot b = [\varphi(e)(b)]a$$

so $\varphi(e)(b) \in Z(A) = R$. Therefore the image of $\varphi(e)$ is R. This completes the proof.

<u>Corollary 3.2</u>: Let A be a central separable R-algebra. If $\mathfrak{a}$ is any ideal of R, then $\mathfrak{a}A \cap R = \mathfrak{a}$.

<u>Proof</u>: By the proposition, $A = L \oplus R$ for some R-submodule L of A. Then

$$\mathfrak{a}A \cap R = \mathfrak{a}(L \oplus R) \cap R = (\mathfrak{a}L \oplus \mathfrak{a}) \cap R = \mathfrak{a}.$$

Proposition 3.3: Let A and B be central separable R-algebras. Then $A \otimes B$ is a central separable R-algebra.

Proof: Since R is an R-direct summand of both A and B, $R \otimes R \simeq R$ can be identified with a subring of $A \otimes B$, so, in particular, $A \otimes B \neq 0$. Therefore $A \otimes B$ is separable by Proposition 1.6.

Now by the hom-tensor relation (1.2.4), we have $\mathrm{Hom}_{A^e}(A,A) \otimes \mathrm{Hom}_{B^e}(B,B) \simeq \mathrm{Hom}_{A^e \otimes B^e}(A \otimes B, A \otimes B) = \mathrm{Hom}_{(A \otimes B)^e}(A \otimes B, A \otimes B)$. But by Corollary 1.3, this is equivalent to

$$A^A \otimes B^B \simeq (A \otimes B)^{A \otimes B}$$

or

$$R \simeq R \otimes R \simeq Z(A \otimes B).$$

Therefore $A \otimes B$ is central.

These propositions are needed to prove the extremely important

Theorem 3.4: The following statements about the R-algebra A are equivalent:

i) A is central separable over R.

ii) A is an A^e-progenerator and A is R-central.

iii) A is an R-progenerator and the map φ from A^e to $\mathrm{Hom}_R(A,A)$ is an isomorphism.

Proof: It is obvious that ii) implies i).

The equivalence of ii) and iii) follows from the Morita Theorems. For example, if A is R-central we have $\mathrm{Hom}_{A^e}(A,A) \simeq R$, so A is an R-progenerator if A is an A^e-progenerator. Moreover, A^e would then be isomorphic by 1.3.4a to $\mathrm{Hom}_R(A,A)$ under left multiplication, which is precisely the map φ, so ii) implies iii). The reverse process takes us from iii) to ii).

The only remaining implication is ii) from i). By definition A is A^e-projective and it is obvious that 1 generates A over A^e, so A is finitely generated over A^e. What remains is to show that A is an A^e-generator and that is more challenging. To accomplish it we must show that

$$A^* \otimes_{\mathrm{Hom}_{A^e}(A,A)=R} A = \mathrm{Hom}_{A^e}(A,A^e) \otimes A$$

is isomorphic to A^e under the map $f \otimes a \to f(a)$. But by Corollary 1.4, $A^* = \mathrm{Hom}_{A^e}(A,A^e)$ is isomorphic to $0:J$ under map $f \to f(1)$. Therefore we have to prove that

$$0:J \otimes A \simeq A^e$$

under $b \otimes a \to (a \otimes 1)b = (1 \otimes a)b$. But this last is easily seen to be equivalent to $A^e(0:J) = A^e$. To obtain this result, we need the following lemma, which we will later improve. (See Corollary 3.7.)

Lemma 3.5: For every maximal two-sided ideal M of a central separable R-algebra A, there exists an ideal $\mathfrak{a}$ of R with $\mathfrak{a}A = M$.

Proof: Let $\mathfrak{a} = M \cap R$ and set $M_o = \mathfrak{a}A$. Clearly $M_o \subseteq M$ and we wish to show equality. By Proposition 1.11 A/M is a central separable algebra over $R/\mathfrak{a}$. But A/M is simple so its center $R/\mathfrak{a}$ is a field. However since $M_o \cap R = \mathfrak{a}$ by Corollary 3.2 A/M_o is also a central separable algebra over the field $R/\mathfrak{a}$ and so is simple by our earlier comment. Therefore M/M_o is 0 or $M = M_o$ and the lemma is proved.

We now complete the proof of Theorem 3.4 by assuming $A^e(0:J)$ is not all of A^e and proceeding to a contradiction. If $A^e(0:J)$ is a proper ideal of A^e, it is contained in some maximal two-sided ideal M of A^e. A^e is a central separable R-algebra by Proposition 3.3. Therefore by the lemma, $M = \mathfrak{a} A^e$ for some proper ideal $\mathfrak{a}$ of R, so

$$A^e(0:J) \subseteq \mathfrak{a}A^e .$$

Applying the multiplication map μ, we get

$$A^e \cdot \mu(O:J) \subseteq \mu(\mathfrak{a}A^e) = \mathfrak{a}A .$$

By Corollary 1.4, $\mu(O:J) = R$, so $A^e \cdot R = A \subseteq \mathfrak{a}A$. But $A = \mathfrak{a}A$ implies $A^e = \mathfrak{a}A^e = M$, a contradiction. Therefore $A^e(O:J) = A^e$ and A is an A^e-generator, concluding the proof.

<u>Corollary 3.6</u>: If A is a central separable R-algebra, then for any R-module M,

$$(M \otimes A)^A \simeq M \qquad \text{as R-modules under the map}$$

$$m \otimes 1 \leftrightarrow m$$

and for any two-sided A/R-module N

$$N^A \otimes A \simeq N \qquad \text{as two-sided A/R-modules under}$$

the map $\Sigma n_i \otimes a_i \leftrightarrow \Sigma n_i a_i$.

<u>Proof</u>: Since A is an R-progenerator and $\mathrm{Hom}_R(A,A) \simeq A^e$, we have by the Morita theorems that $\mathrm{Hom}_{A^e}(A,A^e) \otimes_{A^e} (M \otimes A) \simeq M$. But by the hom-tensor relation (1.2.7),

$$\mathrm{Hom}_{A^e}(A,A^e) \otimes_{A^e} (M \otimes A) \simeq \mathrm{Hom}_{A^e}(A,A^e \otimes_{A^e} (M \otimes A))$$

$$\simeq \mathrm{Hom}_{A^e}(A, M \otimes A) \simeq (M \otimes A)^A, \text{ so } M \simeq (M \otimes A)^A .$$

Also, by a similar argument, $N \simeq (\mathrm{Hom}_{A^e}(A,A^e) \otimes_{A^e} N) \otimes A$ $\simeq \mathrm{Hom}_{A^e}(A,N) \otimes A \simeq N^A \otimes A$. We leave it to the reader to check that in each case the composition of the various isomorphisms gives the map described in the statement of the corollary.

<u>Corollary 3.7</u>: Let A be a central separable R-algebra. Then there is a one-one correspondence between ideals $\mathfrak{a}$ of R and two-sided ideals $\mathfrak{A}$ of A given by $\mathfrak{a} \rightarrow \mathfrak{a}A$ and $\mathfrak{A} \rightarrow \mathfrak{A} \cap R$.

<u>Proof:</u> By Corollary 3.6, $\mathfrak{a} \simeq (\mathfrak{a} \otimes A)^A \simeq (\mathfrak{a}A)^A = \mathfrak{a}A \cap R$ and $\mathfrak{A} \simeq \mathfrak{A}^A \otimes A = (\mathfrak{A} \cap R) \otimes A \simeq (\mathfrak{A} \cap R)A$. This establishes that the

correspondence is one-one. (The reader should verify that the composition of the isomorphisms is equality in each case.)

We conclude this section by proving a result that justifies our division of the study of separable algebras into the central and commutative cases.

Theorem 3.8: An R-algebra A is separable if and only if A is separable as an algebra over its center and its center is a separable R-algebra.

Proof: It follows by the transitivity of separability (Proposition 1.12) that if A is separable over its center C and C is R-separable, then A is R-separable.

Conversely, assuming A is R-separable, we know that A is separable when considered as an algebra over its center. It remains to show that C is R-separable. Since A is central separable over C, A and A° are C-projective by Theorem 3.4. It follows (for example, by an easy argument involving dual bases) that $A \otimes A^{\circ}$ is projective over $C \otimes C$. Also A is projective over $A \otimes A^{\circ}$. Combining these two facts gives us that A is projective as a $C \otimes C$-module by Proposition 1.1.4, so any $C \otimes C$-direct summand of A is a projective $C \otimes C$-module. However, because C is the center of A, any C-direct summand of A is a $C \otimes C$-direct summand and by Proposition 3.1 C is a C-direct summand of A. Hence C is $C \otimes C$-projective, that is, C is R-separable.

§ 4. The Commutator Theorems

In this section we will prove two striking theorems on the representation of a central separable algebra as the tensor product of subalgebras.

To begin, we remark that when A is a central separable R-algebra, we have seen that A is an R-progenerator and that $\mathrm{Hom}_R(A,A)$, being isomorphic to $A \otimes A^{\circ}$, is a central separable R-algebra. We have also seen by Example II that $\mathrm{Hom}_R(E,E)$ is a (central) separable R-algebra when E is a finitely generated, free R-module. The first proposition generalizes both of these facts.

<u>Proposition 4.1</u>: Let E be any R-progenerator. Then $A = \mathrm{Hom}_R(E,E)$ is a central separable R-algebra.

<u>Proof</u>: We will show that A satisfies the conditions of iii) of Theorem 3.4. By the corollary to the Dual Basis Lemma, $E^* \otimes E \simeq \mathrm{Hom}_R(E,E) = A$. But E^* is finitely generated and projective by the Morita Theorems, so $A \simeq E^* \otimes E$ is finitely generated and projective, since it is the tensor product of two finitely generated and projective modules. Moreover, it is clear that A is R-faithful since E is R-faithful. Hence A is an R-progenerator by 1.1.9. Finally we must show that the map α from A^e to $\mathrm{Hom}_R(A,A)$ is an isomorphism. By Morita, we have $A^\circ \simeq \mathrm{Hom}_R(E^*,E^*)$. Therefore, using 1.2.6 we have

$$A \otimes A^\circ \simeq \mathrm{Hom}_R(E,E) \otimes \mathrm{Hom}_R(E^*,E^*) \simeq \mathrm{Hom}_R(E \otimes E^*, E \otimes E^*) \simeq \mathrm{Hom}_R(A,A).$$

The reader may laboriously check that the composition of these isomorphisms is indeed α.

<u>Corollary 4.2</u>: Let A be an R-algebra which is a progenerator as an R-module. Then R is an R-direct summand of any subalgebra B of A.

<u>Proof</u>: By the proposition $\mathrm{Hom}_R(A,A)$ is a central separable R-algebra. A can be considered a subalgebra of $\mathrm{Hom}_R(A,A)$ by identifying it with the set of left multiplications by elements of A, under which identification we have $R \subseteq B \subseteq A \subseteq \mathrm{Hom}_R(A,A)$. But by Lemma 3.1, R is an R-direct summand of $\mathrm{Hom}_R(A,A)$, that is, there exists an R-submodule L of $\mathrm{Hom}_R(A,A)$ with $R \cap L = (0)$ and $R + L = \mathrm{Hom}_R(A,A)$. Then $R \cap (B \cap L) = (0)$ and $R + (B \cap L) = B$, so R is an R-direct summand of B.

Consider now the following general situation. Let A be any R-algebra and B any R-subalgebra of A. Then A is naturally a two-sided B/R-module and, as before, we let $A^B = \{a \in A | ab = ba, \forall b \in B\}$. In this setting A^B is seen to be an R-subalgebra of A which commutes with B.

<u>Theorem 4.3</u>: Let A be a central separable R-algebra. Suppose B is any separable subalgebra of A containing R. Set $C = A^B$. Then C is a separable subalgebra of A and $A^C = B$. If B is also central, so is C and the R-algebra map $B \otimes C \to A$ given by $b \otimes c \to bc$ is an isomorphism.

<u>Proof</u>: I). We begin by proving the theorem under the assumption that B is central as well as separable. Then $B \otimes C$ is isomorphic to A under the given map by Corollary 3.6. (Note that the map is a ring homomorphism because B and C commute). Since R is an R-direct summand of B, C is R-separable by Corollary 1.9. Moreover, if x is in the center of C, it commutes with all of $B \otimes C \cong A$. Therefore x is in the center of A which is R, showing that C is central separable. Finally, since C is central separable, we have by 3.6 that $A^C = (B \otimes C)^C = B$.

II). Secondly, we prove the theorem in the case that $A = Hom_R(M,M)$ for some R-progenerator M. In this case, if B is an arbitrary separable subalgebra of A containing R, we have $C = A^B = [Hom_R(M,M)]^B = Hom_B(M,M)$. Now by 3.8 B is central separable over its center Z(B) and Z(B) is separable over R. Moreover the lifting property for projectives (2.3) allows us to conclude that M is a Z(B)-progenerator. (M is Z(B)-faithful because it is A-faithful). Hence B can be considered as a central separable subalgebra of $Hom_{Z(B)}(M,M)$, which itself is central separable over Z(B) by Proposition 4.1. Applying the results of the preceding paragraph, we have $C = A^B = Hom_B(M,M) = [Hom_{Z(B)}(M,M)]^B$ is a central separable Z(B)-algebra and $[Hom_{Z(B)}(M,M)]^C = B$. Now transitivity of separability (1.12) yields that C is a separable R-algebra and the fact that $Z(B) \subseteq C$ implies $B = [\mathrm{Hom}_{Z(B)}(M,M)]^C = [\mathrm{Hom}_R(M,M)]^C = A^C$.

III) We finally treat the general case of the theorem. For any central separable R-algebra A, its enveloping algebra $A \otimes A^\circ$ is a central separable R-algebra (3.3) of the form $\mathrm{Hom}_R(A,A)$ where A is an R-progenerator (3.4), so case II can be applied to the subalgebras of $A \otimes A^\circ$.

We first remark that since both A and A° are R-projective and faithful, A and A° can be regarded as subrings of $A \otimes A^\circ$ and $(A \otimes A^\circ)^{A^\circ} = A$ by Case I. Therefore for any subalgebra A' of A, the subalgebra $A' \otimes A^\circ$ of $A \otimes A^\circ$ has the property

$$(A \otimes A^\circ)^{A' \otimes A^\circ} = ((A \otimes A^\circ)^{A^\circ})^{A'} = A^{A'}. \qquad (*)$$

Now let B be any separable subalgebra of A containing R. Then $B \otimes A^\circ$ is a separable subalgebra of $A \otimes A^\circ$, so applying case II and (*), we have $(A \otimes A^\circ)^{B \otimes A^\circ} = A^B = C$ is a separable R-algebra and

$$B \otimes A^\circ = (A \otimes A^\circ)^{(A \otimes A^\circ)^{B \otimes A^\circ}} = (A \otimes A^\circ)^C .$$

It follows that $A \cap (B \otimes A^\circ) = A \cap (A \otimes A^\circ)^C = A^C$, so the proof is complete if we show $A \cap (B \otimes A^\circ) = B$. However, since $A^\circ = R \oplus L$ for some R-submodule L of A°, $B \otimes A^\circ \simeq B \oplus (B \otimes L)$ and $A \otimes A^\circ \simeq A \oplus (A \otimes L)$, so $(B \otimes L) \cap A \subseteq (A \otimes L) \cap A = (0)$. Consequently any element of $B \oplus (B \otimes L)$ in A must be in B.

While the foregoing theorem states that the central separable subalgebras of a central separable algebra A occur in pairs, each member of a pair being the commutator subalgebra of the other and whose tensor product is isomorphic to A, the next tells us that any representation of A as a tensor product of subalgebras can only occur in this way.

<u>Theorem 4.4</u>: Let A be a central separable R-algebra. Suppose B and C are subalgebras such that the map from $B \otimes C$ to A given by

$b \otimes c \to bc$ is an R-algebra isomorphism. Then B and C are central separable R-algebras with $A^B = C$ and $A^C = B$.

Proof: We know by 4.2 that R is an R-direct summand of both B and C, so B and C are separable algebras by 1.9. Since elements of B commute with elements of C ($(b \otimes 1)(1 \otimes c) = b \otimes c = (1 \otimes c)(b \otimes 1)$), the center of B is contained in the center of $B \otimes C \simeq A$, so B is central as well as separable. Similarly for C. Finally, by 3.6, $A^B = (B \otimes C)^B = C$ and $A^C = (B \otimes C)^C = B$. This concludes the proof.

§ 5. The Brauer Group

We devote this section to the definition and fundamental properties of the Brauer group of a commutative ring. For any commutative ring R, this abelian group reflects the complexity and variety of the central separable R-algebras and is an important invariant of the ring.

For any commutative ring R, consider a collection $\mathfrak{S}(R)$ of central separable R-algebras such that every central separable R-algebra is isomorphic as an R-algebra to exactly one member of $\mathfrak{S}(R)$. To see that $\mathfrak{S}(R)$ is a set, we observe that any element A in $\mathfrak{S}(R)$ is a finitely generated R-module and so is a homomorphic image of $R^{(n)}$ for some integer n. Thus up to R-module isomorphism there is only a set of finitely generated R-modules. But the algebra structure of A is determined by a mapping from $A \otimes A^{o}$ to A and the collection of all such maps is a set. Hence for each isomorphism class of finitely generated R-modules there is only a set of algebra structures which can be given to a representative of that class, and so $\mathfrak{S}(R)$ is a set, as it is the union of sets over an indexing set.*

We can put a commutative, associative, binary operation on $\mathfrak{S}(R)$ by identifying $A \otimes B$ with the element of $\mathfrak{S}(R)$ to which it is isomorphic, where A and B are any two elements of $\mathfrak{S}(R)$. Since $\mathfrak{S}(R)$ contains an

---

* We have been assured that there are ways of forming the collection $\mathfrak{S}(R)$ that are acceptable within the Gödel-Bernays system.

element isomorphic with R, $\mathfrak{S}(R)$ possesses an identity for this operation and therefore forms a commutative monoid under $\otimes$.

For any R-progenerator E, we know by Proposition 4.1 that $\mathrm{Hom}_R(E,E)$ is a central separable R-algebra. Let $\mathfrak{S}^{\circ}(R)$ equal the subset of $\mathfrak{S}(R)$ consisting of those central separable R-algebras A such that $A \cong \mathrm{Hom}_R(E,E)$ for some progenerator E. If E_1 and E_2 are progenerators, so is $E_1 \otimes E_2$ (1.2.3) and the hom-tensor relation (1.2.6) gives $\mathrm{Hom}_R(E_1 \otimes E_2, E_1 \otimes E_2) \cong \mathrm{Hom}_R(E_1,E_1) \otimes \mathrm{Hom}_R(E_2,E_2)$, so $\mathfrak{S}^{\circ}(R)$ is closed under tensor product. Furthermore, since R is an R-progenerator with $R \cong \mathrm{Hom}_R(R,R)$, $\mathfrak{S}^{\circ}(R)$ contains the identity of $\mathfrak{S}(R)$ and is a submonoid of $\mathfrak{S}(R)$.

We introduce a relation $\sim$ in $\mathfrak{S}(R)$ by specifying that two elements A and B of $\mathfrak{S}(R)$ are in relation (written $A \sim B$) if and only if there exist elements Y and Z of $\mathfrak{S}^{\circ}(R)$ such that $A \otimes Y \cong B \otimes Z$ as R-algebras, that is, if and only if there exist R-progenerators E_1 and E_2 such that

$$A \otimes \mathrm{Hom}_R(E_1,E_1) \cong B \otimes \mathrm{Hom}_R(E_2,E_2).$$

It follows that because $\mathfrak{S}^{\circ}(R)$ is a submonoid of $\mathfrak{S}(R)$ the relation $\sim$ is an equivalence relation.

<u>Definition of the Brauer group</u>: Let B(R) denote the equivalence classes of $\mathfrak{S}(R)$ under the relation $\sim$ and let [A] denote the class containing A. We define a binary operation in B(R) by $[A][B] = [A \otimes B]$. If $A' \in [A]$ and $B' \in [B]$, by definition there are Y, Y', Z, Z' in $\mathfrak{S}^{\circ}(R)$ such that $A \otimes Y \cong A' \otimes Y'$ and $B \otimes Z \cong B' \otimes Z'$, from which we obtain

$$(A \otimes B) \otimes (Y \otimes Z) \cong (A \otimes Y) \otimes (B \otimes Z) \cong (A' \otimes Y') \otimes (B' \otimes Z') \cong (A' \otimes B') \otimes (Y' \otimes Z').$$

Thus the operation is well-defined. Obviously it is commutative, associative, and possesses an identity [R]. Finally by Theorem 3.4 A is an R-progenerator and $A \otimes A^{\circ}$ is isomorphic to $\mathrm{Hom}_R(A,A)$ in $\mathfrak{S}^{\circ}(R)$, whence $[A][A^{\circ}] = [\mathrm{Hom}_R(A,A)] = [\mathrm{Hom}_R(R,R)] = [R]$, proving that $[A]^{-1} = [A^{\circ}]$. Thus B(R) forms an abelian group, known as the <u>Brauer group of R</u>.

In case R is a field, B(R) is in one-one correspondence with the isomorphism classes of division algebras having R as center and which are finite dimensional over R. To see this we first observe that since a central separable algebra A over a field is simple of finite dimension over its center, it is isomorphic by the Wedderburn theorem to the full ring D_n of n by n matrices with entries from an R-central division algebra D. Then $A \otimes R \cong A \cong D_n \cong D \otimes R_n$, so [A] = [D] in B(R). (Notice that for a field R, $\mathfrak{S}^{\circ}(R)$ consists of the matrix rings R_n over R, for $n = 1,2,3,\cdots$.) Moreover, if D_1 and D_2 are finite dimensional division algebras having R as center and $[D_1] = [D_2]$, then there exist positive integers n_1 and n_2 such that $(D_1)_{n_1} \cong D_1 \otimes R_{n_1} \cong D_2 \otimes R_{n_2} \cong (D_2)_{n_2}$, from which it follows that $n_1 = n_2$ and $D_1 \cong D_2$. [See Cor. 1.3 of Chap. V.] Thus [A] = [B] if and only if the associated division algebras are isomorphic.

We next establish the basic functorial properties of the Brauer group.

Lemma 5.1: Let A be a central separable R-algebra. Then for any commutative R-algebra S, $A \otimes S$ is a central separable S-algebra.

Proof: We know that $A \otimes S$ contains a copy of S since R is an R-direct summand of A. Thus $A \otimes S \neq 0$ and is therefore separable over S by Corollary 1.7. Moreover, by the hom-tensor relation (1.2.5),

$$R\otimes S \cong \operatorname{Hom}_{A^e}(A,A) \otimes S \cong \operatorname{Hom}_{A^e\otimes S}(A\otimes S, A\otimes S) \cong \operatorname{Hom}_{(A\otimes S)\otimes_S(A\otimes S)^{\circ}}(A\otimes S, A\otimes S)$$

so by 1.3 we have $S \cong R \otimes S$ is the center of $A \otimes S$.

Let $f:R \to S$ be any ring homomorphism of R to the commutative ring S. Then f endows S with a structure as an R-algebra. By Lemma 5.1, $A \in \mathfrak{S}(R)$ implies $A \otimes S$ is isomorphic to an element of $\mathfrak{S}(S)$, so f induces a map from $\mathfrak{S}(R)$ to $\mathfrak{S}(S)$. Further, if E is an R-progenerator, $E \otimes S$ is an S-progenerator and the hom-tensor relation 1.2.5 gives

$\mathrm{Hom}_R(E,E) \otimes S \cong \mathrm{Hom}_S(E\otimes S, E\otimes S)$. Thus $\mathfrak{S}^\circ(R)$ is taken into $\mathfrak{S}^\circ(S)$ under this mapping of $\mathfrak{S}(R)$ into $\mathfrak{S}(S)$. It follows that the map $B(f)\colon B(R) \to B(S)$ given by $B(f)([A]) = [A\otimes S]$ is well-defined, that is, independent of the choice of representative in $[A]$. Also, for A, B in $\mathfrak{S}(R)$, $(A\otimes S)\otimes_S(B\otimes S) \cong (A\otimes B)\otimes S$, so $B(f)$ is a group homomorphism.

Suppose $g\colon S \to T$ is also a ring homomorphism from S to the commutative ring T. Then T is an S-algebra via g and an R-algebra via gf in such a way that for any A in $\mathfrak{S}(R)$, $(A\otimes S)\otimes_S T \cong A\otimes T$, so $B(g)B(f) = B(gf)$. This along with the obvious remark that the identity on R induces the identity on $B(R)$ proves

<u>Proposition 5.2</u>: $B(\ )$ is a covariant functor from the category of commutative rings (and ring homomorphisms) to the category of abelian groups (and group homomorphisms).

It is apparent from the definition of the equivalence "$\sim$" of two elements of $\mathfrak{S}(R)$ that every element of $\mathfrak{S}^\circ(R)$ is equivalent to R. The next proposition shows that conversely $A \sim R$ implies that A is in $\mathfrak{S}^\circ(R)$.

<u>Proposition 5.3</u>: For every A in $\mathfrak{S}(R)$, $A \sim R$ if and only if $A \in \mathfrak{S}^\circ(R)$.

<u>Proof</u>: If $A \sim R$, then there exist R-progenerators E_1 and E_2 such that $A\otimes \mathrm{Hom}_R(E_1,E_1) \cong R\otimes \mathrm{Hom}_R(E_2,E_2) \cong \mathrm{Hom}_R(E_2,E_2)$. Setting $B = \mathrm{Hom}_R(E_1,E_1)$, we can identify A and B via this isomorphism with subalgebras of $\mathrm{Hom}_R(E_2,E_2)$ with the result that $A\otimes B \cong A\cdot B = \mathrm{Hom}_R(E_2,E_2)$. E_2 can then be viewed as a left B-module under the operation $b\cdot e = b(e)$, for $b \in B$, $e \in E_2$. Since $B = \mathrm{Hom}_R(E_1,E_1)$ with E_1 an R-progenerator, we have by the Morita theorem that the R-module $M = E_1^*\otimes_B E_2$ corresponds to E_2 under the category isomorphism between the category of R-modules and the category of left B-modules. It follows from 1.3.1 that $\mathrm{Hom}_R(M,M) \cong \mathrm{Hom}_B(E_2,E_2)$ as R-algebras. But $\mathrm{Hom}_B(E_2,E_2) = \{f \in \mathrm{Hom}_R(E_2,E_2) \mid f(b(e)) = b(f(e)), \text{ for all } b \in B, e \in E_2\}$

$= [\mathrm{Hom}_R(E_2,E_2)]^B \cong A$ by Theorem 4.4, yielding $A \cong \mathrm{Hom}_R(M,M)$. This will establish that A is in $\mathfrak{S}^\circ(R)$ once we verify that M is an R-progenerator.

Since E_2 is R-projective, it follows from Proposition 2.3 that E_2 is B-projective. Because projectivity is preserved under category isomorphism, M (being the image of E_2 under the category isomorphism from ${}_B\mathfrak{M}$ to ${}_R\mathfrak{M}$ induced by E_1) is R-projective. In addition, E_1^* and E_2 are both finitely generated over R, so $E_1^* \otimes E_2$ is, implying that $M = E_1^* \otimes_B E_2$, as a homomorphic image of $E_1^* \otimes E_2$, is finitely generated. Finally because $\mathrm{Hom}_R(M,M) \cong A$ is a faithful R-algebra, M must be a faithful R-module. Therefore M is an R-progenerator by 1.1.9.

<u>Corollary 5.4</u>: Let A and B be central separable R-algebras. Then $A \sim B$ if and only if $A \otimes B^\circ \cong \mathrm{Hom}_R(E,E)$ for some R-progenerator E.

<u>Proof</u>: $A \sim B$ if and only if $[A] = [B]$ if and only if $[A][B]^{-1} = [R]$ if and only if $[A \otimes B^\circ] = [R]$. Now apply the proposition.

Let A be a central separable R-algebra. Any commutative R-algebra S such that $A \otimes S \cong \mathrm{Hom}_S(E,E)$ for some S-progenerator E is called a <u>splitting ring</u> of A and we say that S <u>splits</u> A. For any commutative R-algebra S the map f from R to S given by $f(r) = r \cdot 1$ is a ring homomorphism, inducing the map B(f) from B(R) to B(S). The kernel of B(f) is a subgroup of B(R), denoted <u>B(S/R)</u>, which consists of those classes [A] of B(R) such that $[A \otimes S] = [S]$ in B(S). Thus by Proposition 5.3 [A] is in B(S/R) if and only if S is a splitting ring for A.

It is quite easy to see (Exercise 9) that if $\mathfrak{m}$ is any maximal ideal of R, "the" algebraic closure S of the field $R/\mathfrak{m}$ is an R-algebra which splits any central separable R-algebra. However the placing of the mildest sort of conditions on the splitting ring makes the problem of existence of splitting rings apparently much more difficult. For instance, if we define an <u>extension</u> of R to be a faithful commutative R-algebra, it remains an open question whether or not every central separable R-algebra is split by some extension of R.

A commutative subalgebra S of a central separable R-algebra A is called a <u>maximal commutative subalgebra</u> if it is contained in no larger commutative subalgebra of A. An easy application of Zorn's lemma indicates that maximal commutative subalgebras always exist. Moreover one can see after a moment's reflection that $A^S = \{a \in A | as = sa$, for all $s \in S\}$ always contains S and equals S if and only if S is maximal. In the classical setting, that is, when R is a field, the importance of the maximal commutative subalgebras of central simple R-algebras is their connection with splitting rings (fields) and thereby with the theory of crossed products and Galois cohomology (see Artin, Nesbitt, and Thrall [8], as well as Chapter IV). This connection in the general setting of commutative rings is given by the next theorem.

<u>Theorem 5.5</u>: Let S be a separable extension of R which is an R-progenerator. Then S splits the central separable R-algebra A if and only if there is a central separable R-algebra B in [A] such that B contains a maximal commutative subalgebra isomorphic to S.

<u>Proof</u>: Let us begin by supposing that A is a central separable R-algebra split by S. Since B(S/R) is a subgroup, A° is also split by S, so there exists an S-progenerator E such that $S \otimes A^\circ \cong Hom_S(E,E)$. Because S is an R-progenerator, we have by 1.1.6 that E is an R-progenerator, so that $Hom_S(E,E)$ is contained in the central separable R-algebra $Hom_R(E,E)$. Moreover since both S and A° are R-projective and hence flat, S and A° can be identified with commuting subrings of $Hom_R(E,E)$ so that $S \cdot A^\circ = Hom_S(E,E)$.

Now set $B = [Hom_R(E,E)]^{A^\circ}$. By Theorem 4.3 B is a central separable R-algebra with $B \otimes A^\circ \cong Hom_R(E,E)$. Applying Corollary 5.4, we conclude that B is in [A]. It remains to show that S is a maximal commutative subalgebra of B.

Clearly $S \subseteq [Hom_R(E,E)]^{A^\circ} = B$ since S and A° are commuting subalgebras of $Hom_R(E,E)$. Furthermore, any x in B^S has the property that

x is in $[\mathrm{Hom}_R(E,E)]^S = \mathrm{Hom}_S(E,E) = S\cdot A^{\circ} \cong S \otimes A^{\circ}$. But being in B, x commutes with elements of A° as well as with those of S and so lies in the center S of $S \otimes A^{\circ}$. Therefore $B^S = S$ and S is a maximal commutative subalgebra of B.

Conversely assume S is a maximal commutative subalgebra of a central separable R-algebra B. B° is R-projective, so $S \otimes B^{\circ}$ can be identified with a subalgebra of $B \otimes B^{\circ}$. Moreover since S is R-separable and B is an R-progenerator, B is an S-progenerator (Proposition 2.3). Therefore $B \otimes B^{\circ}$ is an $S \otimes B^{\circ}$-progenerator. (Lemma 1.2.2) But B is a $B \otimes B^{\circ}$-progenerator (Theorem 3.4), so the transitive property of progenerators (Lemma 1.1.6) implies that B is an $S \otimes B^{\circ}$-progenerator.

By Theorem 3.4 $B \otimes B^{\circ} \cong \mathrm{Hom}_R(B,B)$ from which we obtain

$$\begin{aligned}
\mathrm{Hom}_{S\otimes B^{\circ}}(B,B) &= [\mathrm{Hom}_R(B,B)]^{S\otimes B^{\circ}} \\
&\cong (B \otimes B^{\circ})^{S\otimes B^{\circ}} \\
&= ((B \otimes B^{\circ})^{B^{\circ}})^{S} \\
&\cong B^S \qquad \text{(by Theorem 4.3)} \\
&= S
\end{aligned}$$

since S is a maximal commutative subalgebra of B. Thus we have shown that

1) B is an $S \otimes B^{\circ}$-progenerator

and 2) $\mathrm{Hom}_{S\otimes B^{\circ}}(B,B) \cong S$.

It follows from an application of the Morita theorem that $\mathrm{Hom}_S(B,B) \cong S \otimes B^{\circ}$. Since we have already confirmed that B is an S-progenerator, we have proved that B° and therefore B is split by S. This proves the theorem.

It should be remarked that in the proof of Theorem 5.4 we needed only that S is an R-progenerator to prove that if S splits A, S is a maximal commutative subalgebra of some element of [A], while it is sufficient that S be separable to prove the converse.

In order to apply Theorem 5.5 we need to know that there are splitting rings which are separable as algebras and progenerators as modules over R. At this time the existence of splitting rings having these properties is an open question for a general commutative ring R. However the interested reader may consult Auslander-Goldman [7] and Endo-Watanabe [39] for a proof that such splitting rings exist when R is a local ring.

§ 6. Automorphisms of Central Separable Algebras

We now present general results on the nature of the automorphisms of a central separable algebra. The first theorem, due to Rosenberg and Zelinsky [83], relates the group of automorphisms of a central separable R-algebra to the class group P(R) of R. Also, for a central separable R-algebra of the form $\mathrm{Hom}_A(M,M)$ where A is itself a central separable R-algebra and M is an A-progenerator, we obtain a description (due to Rosenberg and Zelinsky [83] as modified by Knus [67]) of the isomorphism classes of A-progenerators N with $\mathrm{Hom}_A(N,N) \cong \mathrm{Hom}_A(M,M)$ as R-algebras.

We begin with a proposition which is really a corollary to Theorem 4.3. In the proof we use localization techniques for the first time.

Proposition 6.1: Every algebra endomorphism of a central separable algebra is an automorphism.

Proof: Let f be any algebra endomorphism of a central separable algebra A. The kernel $\mathfrak{A}$ of f is a two-sided ideal of A and is therefore of the form $\mathfrak{a}A$ for some ideal $\mathfrak{a}$ of R. (Corollary 3.7) But f takes 1 to 1, so $0 = f(\alpha) = f(\alpha\cdot 1) = \alpha f(1) = \alpha$ for all α in $\mathfrak{a}$, so $\mathfrak{a} = 0$ and f is one-one.

Because f is one-one, f(A) is a central separable subalgebra of A. Setting $C = A^{f(A)}$, we have by Theorem 4.3 that C is central separable and $f(A) \otimes C \cong A$. It follows as in the proof of Theorem 4.3 that f(A) is an R-direct summand of A, that is, A contains an R-submodule L with $A = f(A) \oplus L$. Since A, f(A) and L are all finitely generated and projective over R, $A \otimes R_m$, $f(A) \otimes R_m$, and $L \otimes R_m$ are all free of finite rank over the local ring R_m (for any maximal ideal m of R). But $A \otimes R_m \cong f(A) \otimes R_m$ because f is one-one, so they have the same rank. Since $A \otimes R_m \cong (f(A) \otimes R_m) \oplus (L \otimes R_m)$, the rank of $L \otimes R_m$ is zero, that is, $L \otimes R_m = 0$ for every maximal ideal m of R. Therefore $L = 0$ (1.4.4), so $f(A) = A$ and f is onto.

Let A be a central separable R-algebra. We denote by Aut(A) the group of all R-algebra automorphisms of A. For a unit u in A, the inner automorphism σ_u: $a \to uau^{-1}$ has the property that $\rho\sigma_u\rho^{-1} = \sigma_{\rho(u)}$ for an arbitrary R-automorphism ρ of A, so the set of inner automorphisms forms a normal subgroup Inn(A) of Aut(A).

We can use any R-automorphism σ of A to give A the structure of a left A^e-module by defining $a \otimes a' \cdot b = ab\sigma(a')$. Call this module A_σ. For $\rho, \sigma \in \mathrm{Aut}(A)$, make $A_\rho \otimes_A A_\sigma$ a left A^e-module by $(a \otimes a') \cdot (b \otimes b') = ab \otimes b'\sigma(a')$.

We first show that u: $A_\rho \otimes_A A_\sigma \to A_{\rho\sigma}$ given by $\mu(b \otimes b') = b\rho(b')$ is a left A^e-module isomorphism. We note that since A_σ and $A_{\rho\sigma}$ are isomorphic to A as left A-modules, μ is an isomorphism of abelian groups. Moreover, for any $a \otimes a'$ in A^e, $(a \otimes a') \cdot \mu(b \otimes b') = (a \otimes a') \cdot b\rho(b') = ab\rho(b')\rho\sigma(a') = ab\rho(b'\sigma(a')) = \mu(ab \otimes b'\sigma(a')) = \mu((a \otimes a') \cdot (b \otimes b'))$.

Next consider $(A_\sigma)^A = \{b \in A | ab = b\sigma(a), \text{ for all } a \text{ in } A\}$. We know by Corollary 3.6 that $(A_\sigma)^A \otimes A \cong A_\sigma$ as left A^e-modules. Since R is an R-direct summand of A, it follows that $(A_\sigma)^A$ is an R-direct summand of A_σ and so is finitely generated and projective as an R-module. Furthermore, for any maximal ideal m of R, $A_\sigma \otimes R_m \cong ((A_\sigma)^A \otimes A) \otimes R_m$

$\cong ((A_\sigma)^A \otimes R_m) \otimes_{R_m} (A \otimes R_m)$ and $\text{rank}_{R_m} (A_\sigma \otimes R_m) = \text{rank}_{R_m} (A \otimes R_m)$, so $\text{rank}_{R_m} [(A_\sigma)^A \otimes R_m] = 1$. Therefore $(A_\sigma)^A$ is of rank 1 and determines an element of $P(R)$, the class group of R. Thus we have a map $\alpha: \sigma \to |(A_\sigma)^A|$ of Aut(A) into P(R). By Corollary 3.6 and the preceding paragraph, $[(A_\rho)^A \otimes (A_\sigma)^A] \otimes A \cong [(A_\rho)^A \otimes A] \otimes_A [(A_\sigma)^A \otimes A] \cong A_\rho \otimes_A A_\sigma \cong A_{\rho\sigma} \cong (A_{\rho\sigma})^A \otimes A$, so $(A_\rho)^A \otimes (A_\sigma)^A \cong (A_{\rho\sigma})^A$ as R-modules. Therefore α is a group homomorphism.

<u>Theorem 6.2</u>: The sequence

$$1 \to \text{Inn}(A) \to \text{Aut}(A) \overset{\alpha}{\to} P(R)$$

is exact.

<u>Proof</u>: We need only show that the kernel of α is exactly Inn(A), that is, $(A_\sigma)^A \cong R$ if and only if $\sigma = \sigma_u$ for some unit u. Assuming $\sigma = \sigma_u$, then $(A_\sigma)^A = \{b \in A \mid ab = buau^{-1}$, for all a in A$\} = \{b \in A \mid a(bu) = (bu)a$, for all a in A$\}$, so that $bu \in R$ and $(A_\sigma)^A \cong R$ under the map $b \to bu$. Conversely, if $(A_\sigma)^A \cong R$, then $(A_\sigma)^A$ is a free R-module on one generator, call it v. Since $(A_\sigma)^A \otimes A \cong A_\sigma$ under the map $rv \otimes b \to rvb$ (Corollary 3.6), we see that $vc = 1$ for some c in A. Hence v is a unit of A in $(A_\sigma)^A$, so $av = v\sigma(a)$ or $\sigma(a) = v^{-1}av$ for all a in A. Thus, setting $u = v^{-1}$, we have $\sigma = \sigma_u$.

<u>Corollary 6.3</u>: The group Aut(A)/Inn(A) is abelian and if $P(R) = 1$, every R-automorphism of a central separable R-algebra is inner.

M. A. Knus [67] has generalized this corollary by showing that all automorphisms of all central separable R-algebras are inner if and only if P(R) is torsion-free. More generally, he has given in [66] conditions under which isomorphic central separable subalgebras of a central separable algebra are conjugate under an inner automorphism.

Next suppose A is any central separable R-algebra and M is any A-progenerator. Then M is an R-progenerator (because A is) and $\mathrm{Hom}_R(M,M)$ is a central separable R-algebra. A can be viewed as a central separable subalgebra of $\mathrm{Hom}_R(M,M)$ by identifying an element a of A with scalar multiplication by a in M. Then by Theorem 4.3, $[\mathrm{Hom}_R(M,M)]^A = \mathrm{Hom}_A(M,M)$ is a central separable R-algebra with $A \otimes \mathrm{Hom}_A(M,M) \cong \mathrm{Hom}_R(M,M)$. It is reasonable to wonder whether there are other A-progenerators N such that $\mathrm{Hom}_A(M,M) \cong \mathrm{Hom}_A(N,N)$ (as R-algebras). With this in mind, we denote by $\underline{Q_A(M)}$ the left A-module isomorphism classes of A-progenerators N such that $\mathrm{Hom}_A(M,M) \cong \mathrm{Hom}_A(N,N)$.

Lemma 6.4: An A-progenerator N has the property that $\mathrm{Hom}_A(M,M) \cong \mathrm{Hom}_A(N,N)$ (as R-algebras) if and only if $N \cong M \otimes E$ (as left A-modules) for some finitely generated, projective R-module E of rank 1.

Proof: First suppose $N \cong M \otimes E$ with $|E| \in P(R)$. Then by 1.5.1 $\mathrm{Hom}_R(E,E) \cong R$, giving $\mathrm{Hom}_A(N,N) \cong \mathrm{Hom}_{A\otimes R}(M \otimes E,\ M \otimes E) \cong \mathrm{Hom}_A(M,M) \otimes \mathrm{Hom}_R(E,E) \cong \mathrm{Hom}_A(M,M)$.

We prove the converse first in the case A = R. Suppose N is an R-progenerator with $B = \mathrm{Hom}_R(M,M) \cong \mathrm{Hom}_R(N,N)$. This isomorphism induces a structure on N as a left B-module and by 2.3 N is B-projective. By the Morita theorem the category of left B-modules is isomorphic to the category of R-modules, under which isomorphism N corresponds to some projective R-module E with $M \otimes E \cong N$ and $E \cong \mathrm{Hom}_R(M,R) \otimes_B N$. It is obvious from this last isomorphism that e is also finitely generated over R since M and N are. Therefore we will have proved the case A = R if we show E is of rank one.

N, M, and E are all finitely generated, projective R-modules, so for any maximal ideal $\mathfrak{m}$ of R, $N \otimes R_{\mathfrak{m}}$, $M \otimes R_{\mathfrak{m}}$, and $E \otimes R_{\mathfrak{m}}$ are all free over $R_{\mathfrak{m}}$ of finite rank, say n, m, and e respectively. Then me = n because $M \otimes E \cong N$. Now $\mathrm{Hom}_R(M,M) \cong \mathrm{Hom}_R(N,N)$ implies $\mathrm{Hom}_{R_{\mathfrak{m}}}(M \otimes R_{\mathfrak{m}}, M \otimes R_{\mathfrak{m}}) \cong \mathrm{Hom}_{R_{\mathfrak{m}}}(N \otimes R_{\mathfrak{m}},\ N \otimes R_{\mathfrak{m}})$, so

$m^2 = \text{rank}_{R_m}[\text{Hom}_{R_m}(M \otimes R_m, M \otimes R_m)] = \text{rank}_{R_m}[\text{Hom}_{R_m}(N \otimes R_m, N \otimes R_m)] = n^2$.
Therefore $e = 1$ and E is of rank one.

To prove the lemma for general A, we observe that $\text{Hom}_A(M,M) \cong \text{Hom}_A(N,N)$ and Theorem 4.3 imply $\text{Hom}_R(M,M) \cong A \otimes \text{Hom}_A(M,M) \cong A \otimes \text{Hom}_A(N,N) \cong \text{Hom}_R(N,N)$, so the already established case yields $N \cong M \otimes E$.

It is now apparent that $Q_A(M)$ forms a group under the operation $(M \otimes E_1)(M \otimes E_2) = M \otimes (E_1 \otimes E_2)$ and that the mapping $\beta\colon E \to M \otimes E$ induces a group homomorphism of $P(R)$ onto $Q_A(M)$.

Theorem 6.5: Let A be a central separable R-algebra and set $B = \text{Hom}_A(M,M)$ for some A-progenerator M. Then the sequence

$$1 \to \text{Inn}(B) \to \text{Aut}(B) \overset{\alpha}{\to} P(R) \overset{\beta}{\to} Q_A(M) \to 1$$

is exact.

Proof: It is only necessary to show that the kernel of β equals the image of α.

Begin by assuming $|E|$ is an element of $P(R)$ such that there is an A-module isomorphism $\mu\colon M \otimes E \to M$, that is, assume $M \otimes E \in \text{kernel}(\beta)$. Then 1.2.4 gives $B \otimes E \cong \text{Hom}_A(M,M) \otimes \text{Hom}_R(R,E) \cong \text{Hom}_A(M, M \otimes E) \cong \text{Hom}_A(M,M) = B$, where the composition of the isomorphisms is the map ψ given by $[\psi(b \otimes e)](m) = \mu(b(m) \otimes e)$ for all b in B, e in E, m in M. Having checked that ψ is a right B-module isomorphism, one sets $v = \psi^{-1}(1)$ and observes that every element of $B \otimes E$ has a representation of the form vb for some unique b in B. In particular, since $B \otimes E$ is a left B-module, for every b in B, there exists an element we call $\sigma(b)$ in B with $bv = v\sigma(b)$. Clearly the map $\sigma\colon B \to B$ described by $b \to \sigma(b)$ is an R-algebra endomorphism and is therefore by Proposition 6.1 an automorphism. The mapping $b \to bv$ from $B_{\sigma^{-1}}$ to $B \otimes E$ can be seen to be a B^e-module isomorphism, so $E \cong (B \otimes E)^B \cong (B_{\sigma^{-1}})^B$ by 3.6. Hence $|E|$ is in the image of α.

Conversely suppose $|E|$ is in the image of α, that is, $E \cong (B_\sigma)^B$ for some automorphism σ of B. Since $(B_\sigma)^B \subset B = \mathrm{Hom}_A(M,M)$, we have an R-homomorphism η: $(B_\sigma)^B \otimes M \to M$ given by $\eta(f \otimes m) = f(m)$. We show that η is the desired isomorphism of $E \otimes M$ with M.

First, by Corollary 3.6, $(B_\sigma)^B \otimes B \cong B_\sigma$ as B^e-modules under the map $b \otimes b' \to b\sigma(b')$. Hence there exist b_i in $(B_\sigma)^B$ and b_i' in B such that $\sum_i (b_i\sigma(b_i') = 1$ in B_σ, that is, $\sum_i b_i(\sigma(b_i')(m)) = m$, for every $m \in M$. But this says that $\eta(\sum_i[b_i \otimes \sigma(b_i')(m)]) = m$, so η is onto.

Finally, since η is onto and M is R-projective, the sequence

$$0 \to \ker(\eta) \to E \otimes M \to M \to 0$$

splits, so $E \otimes M \cong M \oplus \ker(\eta)$. But for every maximal ideal $\mathfrak{m}$ of R, the $\mathfrak{m}$-rank of $E \otimes M$ equals the $\mathfrak{m}$-rank of M, so that the $\mathfrak{m}$-rank of the kernel of η is 0 for every $\mathfrak{m}$, implying by 1.4.4 that the kernel of η is 0 and η is one-one. This shows $E \otimes M \cong M$, so $|E| \in$ kernel (β), concluding the proof of the theorem.

Corollary 6.6: If $P(R) = 1$ and $B = \mathrm{Hom}_A(M,M)$ for some central separable R-algebra A and some A-progenerator M, then M is uniquely determined up to isomorphism.

§ 7. Two Criteria for Separability

We would like to conclude Chapter II by presenting an important theorem whose proof involves ideas and techniques more advanced than those of the rest of these notes. To do this we will list without proof a number of results needed for the exposition and point out where their proofs can be found.

There is much known about separable algebras over fields (see Chapter X of Curtis and Reiner[G]) and it is usually quite easy to recognize whether or not a finite dimensional algebra over a field is separable. For example, over a perfect field, an algebra is separable if and only if it contains no nilpotent ideals; that is, if and

and only if it is semisimple. The importance of the theorem we are about to state is that it shows that separability of a finitely generated algebra over any commutative ring depends on the separability of certain closely related algebras over fields. More precisely, we know by Corollary 1.7 that if A is a separable algebra over the commutative ring R, then both $A \otimes R_m$ is R_m-separable and A/mA is R/m-separable, where m is any maximal ideal of R. The theorem of this section contains the converse to these results in the case that the algebra is finitely generated.

<u>Theorem 7.1</u>: The following statements concerning a finitely generated algebra A over a commutative ring R are equivalent:

a) A is a separable R-algebra.

b) $A \otimes R_m \cong A_m$ is a separable R_m-algebra for every maximal ideal m of R.

c) A/mA is a separable R/m-algebra for every maximal ideal m of R.

We precede the proof with some definitions and results. A module M over a (not necessarily commutative) ring R is said to be <u>finitely presented</u> if there is an exact sequence

$$0 \to K \to F \to M \to 0$$

of R-modules with F free and both K and F finitely generated. Clearly any finitely presented module is finitely generated and if R is noetherian the two concepts coincide.

The idea of finite presentation provides a link between flat and projective modules:

I. A module is finitely generated and projective if and only if it is flat and finitely presented.

<u>Proof</u>: For an outline of a proof, see exercise 15, page 64 of [E].

A local ring R with maximal ideal $\mathfrak{m}$ is called a Henselian ring if every monic polynomial f in R[x] whose image $\overline{f}$ in $R/\mathfrak{m}[x]$ factors as $\overline{f} = g_0 h_0$ where g_0, h_0 are relatively prime, monic polynomials in $R/\mathfrak{m}[x]$ has a factorization f = gh in R[x] with g, h monic and $\overline{g} = g_0$, $\overline{h} = h_0$, $\overline{g}$, $\overline{h}$ being the natural images of g and h in $R/\mathfrak{m}[x]$.

An R-module M is called faithfully flat if it is flat and $M \otimes_R N = 0$ implies $N = 0$ for every R-module N.

II.) For any local ring R, there exists a local (commutative) R-algebra R^* such that

a.) R^* is a Henselian ring;

b.) R^* is a faithfully flat R-module; and,

c.) $\mathfrak{m}R^*$ is the maximal ideal of R^* with $R^*/\mathfrak{m}R^* \cong R/\mathfrak{m}$.

[Proof. See pages 179-182 of M. Nagata's "Local Rings," Interscience 1962].

Finally we will need a result of Azumaya's concerning algebras over Henselian rings [8, Theorem 24].

III.) Let A be a finitely generated algebra over the Henselian ring R. Let $\mathfrak{A}$ be a two-sided ideal of A and $\overline{e}$ any idempotent in $A/\mathfrak{A}$. Then there exists an idempotent e in A which is mapped onto $\overline{e}$ under the natural map of A to $A/\mathfrak{A}$.

Using I, II, and III, we can now give a proof of Theorem 7.1.

Proof of 7.1. We know already by 1.7 that a.) implies both b.) and c.).

b.) implies a.). By the definition of separability, $A_\mathfrak{m}$ is $(A_\mathfrak{m})^e$-projective and hence flat, where $\mathfrak{m}$ is any maximal ideal of R. Therefore by exercise 1.11, A is A^e-flat. But if $A = Ra_1 + \cdots + Ra_n$, it is clear that J is generated as a left ideal in A^e by $\{a_i \otimes 1 - 1 \otimes a_i \mid i = 1, \ldots, n\}$, so

$$0 \to J \to A^e \xrightarrow{\mu} A \to 0$$

is a finite presentation of A as an A^e-module. Thus by I) A is A^e-projective and therefore separable.

c.) implies a.). Let m be any maximal ideal of R. Since $R_m/mR_m \cong R/m$ we have $A \otimes R_m/m(A \otimes R_m) \cong A \otimes (R_m/mR_m) \cong A \otimes (R/m) \cong A/mA$. Thus in view of the preceding paragraph, we need only prove that c.) implies a.) when R is a local ring.

Let R^* be an Henselian R-algebra as given by II. Since R^* is faithfully flat over R, it is clear by exercise 1.13 that A will be A^e-flat if $A \otimes R^*$ is $(A \otimes R^*)\otimes_{R^*}(A^\circ \otimes R^*)$-flat. Thus, by I, A is R-separable if and only if $A \otimes R^*$ is R^*-separable. Moreover $(A \otimes R^*)/m(A \otimes R^*) \cong A \otimes (R^*/mR^*) \cong A/mA$, so nothing is lost if we assume that R is Henselian. Thus we must prove that if R is a Henselian ring with maximal ideal m and A is a finitely generated R-algebra, then A is R-separable if A/mA is separable over R/m.

In what follows we will be in the situation of the accompanying commutative diagram, where the vertical maps are the natural ring epimorphisms and the horizontal maps are the multiplication homomorphisms.

$$\begin{array}{ccccc} A^e & \xrightarrow{\mu} & A & \longrightarrow & 0 \\ \downarrow & & \downarrow & & \\ (A/mA)^e & \xrightarrow{\overline{\mu}} & A/mA & \longrightarrow & 0 \end{array}$$

Assuming A/mA is separable, we know that $(A/mA)^e$ contains a separability idempotent $\overline{e}$ for A/mA. We assert that $\overline{e}$ can be lifted to an idempotent e in A^e with $\mu(e) = 1$; that is, there exists an idempotent e in A^e mapping onto $\overline{e}$ under the natural map of A^e onto $(A/mA)^e$ with the property that $\mu(e) = 1$. To see this we first lift $\overline{e}$ to an idempotent e' in A^e, which we can do by III. Since $\overline{\mu}(\overline{e}) = 1$ in A/mA, the commutativity of the diagram implies that $\mu(e') \in 1 + mA$. But $mA = \mathrm{rad}(A)$ so $1 + mA$ consists of units in A.

Set $e = [\mu(e')^{-1} \otimes 1]e'[\mu(e') \otimes 1]$. Clearly e is an idempotent mapping onto $\bar{e}$. Moreover, $\mu(e) = [\mu(e')^{-1} \otimes 1]e' \cdot \mu([\mu(e') \otimes 1]) = [\mu(e')^{-1} \otimes 1]$ $e' \cdot \mu(e') = [\mu(e')^{-1} \otimes 1] \cdot \mu(e'^2) = [\mu(e')^{-1} \otimes 1] \cdot \mu(e') = 1$.

Next we see that the sequence $0 \to J \to A^e \to A \to 0$ splits as R-modules with the splitting map $a \to a \otimes 1$ from A to A^e. Therefore the map π: $\lambda \to (\lambda - \mu(\lambda)\otimes 1)$ is an R-module projection of A^e onto J. Consequently we have an R-module homomorphism $\lambda e \to \lambda e - (\mu(\lambda) \otimes 1)e$ (the composition of the maps $(A^e)e \xrightarrow{\text{incl}} A^e \xrightarrow{\pi} J \xrightarrow{(\ )e} Je$) which is a projection onto Je. Thus there is an R-submodule M of $(A^e)e$ with $(A^e)e = M \oplus Je$.

Now one can readily check that the kernel of the natural map from A^e to $(A/mA)^e$ is $m(A^e)$. It follows that $Je \subseteq m(A^e)$ because the image of Je in $(A/mA)^e$ is zero. But then $Je \subseteq m(A^e)e = mJe \oplus mM$, whence $Je = mJe$. Moreover Je is a finitely generated R-module since $(A^e)e$ is, so by Nakayama's lemma (1.1.8), $Je = 0$. Thus e is a separability idempotent for A, proving the theorem.

Exercises

1.) Derivations and Hochschild Cohomology

Let A be an R-algebra and let M be any two-sided A/R-module. An element g of $\text{Hom}_R(A,M)$ is called a derivation if $g(ab) = a \cdot g(b) + g(a) \cdot b$, for all a, b in A. A derivation g is called inner if there exists an element m in M such that $g(a) = a \cdot m - m \cdot a$ for all a in A. Let $Z_R^1(A,M)$ denote the set of all derivations of A into M along with the R-module structure inherited from $\text{Hom}_R(A,M)$. Let $B_R^1(A,M)$ be the subset of all inner derivations. Clearly $B_R^1(A,M)$ is a submodule of $Z_R^1(A,M)$. Set $H_R^1(A,M) = \dfrac{Z_R^1(A,M)}{B_R^1(A,M)}$. $H_R^1(A,M)$ is called the first Hochschild cohomology module of A with coefficients in M.

Theorem: A is R-separable if and only if $H_R^1(A,M) = 0$ for every two-sided A/R-module M. Give a proof of this theorem in the following steps:

a.) Show that if A is R-separable with separability idempotent $e = \sum x_i \otimes y_i$, then for any derivation g of A into M, $g(a) = a\cdot m - m\cdot a$ where $m = \sum_i x_i g(y_i)$.

b.) Prove the converse by using $H_R^1(A,J) = 0$ and considering $\tau \in Z_R^1(A,J)$ given by $\tau(a) = a \otimes 1 - 1 \otimes a$.

2.) Let R be a commutative ring and A an R-algebra. Let L and N be A-modules. By an <u>extension of L by N</u>, we mean a triple (φ, M, η) where M is an A-module, φ is an A-monomorphism from L into M, η is an A-epimorphism of M onto N and $\mathrm{Im}(\varphi) = \mathrm{Ker}(\eta)$; that is,

$$0 \to L \xrightarrow{\varphi} M \xrightarrow{\eta} N \to 0$$

is exact. We call two extensions (φ, M, η) and (φ', M', η') <u>equivalent</u> if there exists an isomorphism $\theta: M \to M'$ with $\varphi' = \theta\varphi$ and $\eta = \eta'\theta$. Let E(N,L) denote the collection of equivalence classes of extensions of L by N.

An extension (φ, M, η) of L by N is said to be an <u>R-split extension</u> if $\varphi(L)$ is an R-direct summand of M. Let $E^s(N,L)$ denote the collection of equivalence classes of R-split extensions.

$\mathrm{Hom}_R(N,L)$ is a two-sided A/R-module by $(a\cdot f)(n) = af(n)$ and $(f\cdot b)(n) = f(bn)$ for a,b in A, f in $\mathrm{Hom}_R(N,L)$.

Prove that there exists a one-one correspondence between $H_R^1(A, \mathrm{Hom}_R(N,L))$ and $E^s(N,L)$ which corresponds the zero element of $H_R^1(A, \mathrm{Hom}_R(N,L))$ to the equivalence class of extensions containing

$$0 \to L \to L \oplus N \to N \to 0 .$$

Thus if A is a separable R-algebra, every exact sequence of A-modules which R-splits A-splits. This result can be proved directly by the methods of 2.3.

3.) Show that a finite direct sum of separable R-algebras is separable over R.

4.) <u>Question</u>: Is a direct product of an infinite number of separable R-algebras separable?

5.) Give an example of a separable R-algebra which is not finitely generated as an R-module.

6.) a.) Give an example of an algebra A over an arbitrary commutative ring R which is separable and which contains an inseparable subalgebra.

b.) Can you find such an algebra A which is commutative?

7.) (Azumaya) Let A be a finitely generated, faithful R-algebra with generating set $b_1, \cdots, b_n$. Show that A is central separable and free with basis $b_1, \cdots, b_n$ if and only if the matrix (λ_{ij}) with $\lambda_{ij} = b_i b_j$ is invertible in the full matrix ring of n-by-n matrices with entries from A.

8.) Let A be a faithful, finitely generated R-algebra where R is a local ring with maximal ideal m. Show that each of the following conditions is equivalent to A being central separable over R:

i.) A/mA is a central simple R/m-algebra;

ii.) A possesses a generating set $b_1, \cdots, b_n$ over R such that, if $\bar{b}_i$ denotes the image of b_i in A/mA, then the matrix $(\bar{\lambda}_{ij})$ with $\bar{\lambda}_{ij} = \bar{b}_i \bar{b}_j$ is invertible in the ring of all n-by-n matrices with entries from A/mA.

9.) Show that if m is any maximal ideal of a commutative ring R, the algebraic closure S of the field R/m is an R-algebra which splits any central separable R-algebra.

10.) Let A be a finitely generated R-algebra. Show that if A is separable, then rad(A) = $\cap$(mA) where m runs over the maximal ideals of R.

CHAPTER III

This chapter is devoted to a Galois theory of commutative separable algebras which are projective over the coefficient ring. The analogue of a finite separable field extension is a commutative projective separable algebra whose only idempotents are 0 and 1. Proper idempotents are avoided in the Galois theory of commutative rings because they introduce technical difficulties. Throughout this chapter R is a commutative ring and $\otimes$ means tensoring with respect to R.

§1. The Fundamental Theorem

This section develops the basic Galois theory for commutative rings and includes a statement and proof of The Fundamental Theorem of Galois Theory for commutative rings with no idempotents other than 0 and 1.

The ring S is called an *extension* of the commutative ring R in case S is a commutative faithful R-algebra.

If S is an extension of R and H is a set of R-automorphisms of S we let $S^H = \{x \in S \,|\, \sigma(x) = x \text{ for all } \sigma \in H\}$.

If S is an extension of R then we can identify R with $R \cdot 1$ in S and think of R as a subring of S. With this identification, an R-automorphism of S must leave R elementwise fixed; and any automorphism of S leaving R elementwise fixed is an R-automorphism of S. Observe that the set of all R-automorphisms of S forms a group whose operation is composition.

If S is an extension of R and G is the group of all automorphisms of S leaving R elementwise fixed then S is called a *normal extension* of R in case $S^G = R$.

If G is a group and H is a subgroup of finite index, we let $[G:H]$ denote the index of H in G. The order of the finite group G is denoted $[G:1]$. The principal result of this section can now be stated.

Theorem 1.1 (Fundamental Theorem of Galois Theory).
Let S be a finite normal separable extension of R, and assume 0 and 1 are the only idempotents of S. Let G be the group of all automorphisms of S which leave R elementwise fixed. Then G is finite with $[G:1] = \text{Rank}_R(S)$ and there is a one to one correspondence

$$H \to S^H \qquad T \to \{\sigma \varepsilon G \mid \sigma(x) = x \text{ for all } x \varepsilon T\}$$

between the subgroups of G and the subrings of S which contain R and are separable over R. The subgroup H is normal in G if and only if the corresponding subring T is normal over R.

The proof of Theorem 1.1 can be given only after some notions which are important in their own right have been developed. The first step is to show the equivalence of appropriate generalizations of the traditional approaches to the Galois theory of fields. Let G be a finite group of R-automorphisms of the extension S of R. Out of this we want to form two new R-algebras.

First let $\{u_\sigma \mid \sigma \varepsilon G\}$ be a free basis for an S-module. Define multiplication in this module by letting $(au_\sigma)(bu_\tau) = a\sigma(b)u_{\sigma\tau}$ for all $a, b \in S$; $\sigma, \tau \in G$ and extending by linearity. Denote this new R-algebra $\Delta(S:G)$. Define an R-algebra homomorphism $j: \Delta(S:G) \to \text{Hom}_R(S,S)$ by $[j(au_\sigma)]x = a\sigma(x)$.

Form another S-algebra by letting $\{v_\sigma \mid \sigma \varepsilon G\}$ be a free basis for an S-module, and defining multiplication by $(av_\sigma)(bv_\tau) = abv_\sigma \delta_{\sigma\tau}$ for all $a, b \in S$; $\sigma, \tau \in G$. Here $\delta_{\sigma,\tau} = \begin{cases} 0 & \sigma \neq \tau \\ 1 & \sigma = \tau \end{cases}$. Denote this new R-algebra by $\nabla(S:G)$, and notice this is just the direct sum of $[G:1]$ copies of S. View $S \otimes S$ as an S-algebra where S is identified with the first variable. Observe that the map $\ell: S \otimes S \to \nabla(S:G)$ defined by $\ell(a \otimes b) = \sum_{\sigma \varepsilon G} a\sigma(b)\, v_\sigma$ is an S-algebra homomorphism. We can now prove

Proposition 1.2: Let S be an extension of R and let G be a finite group of automorphisms of S, then the following statements are equivalent:

1) i. $S^G = R$

ii. For each non-zero idempotent $e \in S$ and each pair $\sigma \neq \tau$ in G there is an element $x \in S$ with $\sigma(x)e \neq \tau(x)e$

iii. S is a separable R-algebra

2) i. $S^G = R$

ii. There exist $x_1 \ldots x_n$; $y_1 \ldots y_n$ in S with $\Sigma_{j=1}^{n} x_j \, \sigma(y_j) = \delta_{\sigma,1}$ (1 = identity in G)

3) i. S is a finitely generated projective R-module

ii. $j\colon \Delta\ (S:G) \to \mathrm{Hom}_R\ (S,S)$ is an isomorphism

4) i. $S^G = R$

ii. $\ell\colon S \otimes S \to \nabla\ (S:G)$ is an isomorphism

5) i. $S^G = R$

ii. For each maximal ideal M of S and for each $1 \neq \sigma \in G$ there is an $x \in S$ with $(\sigma\ (x) - x) \not\in M$.

__Proof__: 1) implies 2). Since S is separable over R, Proposition 2.1.1 implies the existence of an idempotent $e = \Sigma_{j=1}^{n} x_j \otimes y_j$ in $S \otimes S$ with $\Sigma_{j=1}^{n} x_j y_j = 1$ and $(1 \otimes x - x \otimes 1)e = 0$ for all $x \in S$. If σ is an R-automorphism of S then $1 \otimes \sigma$ is an S-automorphism of $S \otimes S$.

Let $\mu\colon S \otimes S \to S$ be defined by $\mu(x \otimes y) = xy$, since S is commutative μ is an R-algebra homomorphism. Now e is an idempotent in $S \otimes S$, so $e_\sigma = \mu\ (1 \otimes \sigma(e))$ is an idempotent in S. For any $x \in S$

$$\begin{aligned} xe_\sigma &= x\ \mu(1 \otimes \sigma\ (e)) \\ &= \mu\ (x \otimes 1 \cdot 1 \otimes \sigma\ (e)) \\ &= \mu\ (1 \otimes \sigma\ (x \otimes 1 \cdot e)) \\ &= \mu\ (1 \otimes \sigma\ (1 \otimes x \cdot e)) \\ &= \sigma\ (x)\ e_\sigma \end{aligned}$$

therefore $xe_\sigma = \sigma(x)e_\sigma$ for all $x \in S$ so by ii of 1) either $e_\sigma = 0$ or $\sigma = 1$. Therefore

$$\delta_{\sigma,1} = e_\sigma = \mu\ (1 \otimes \sigma(e))$$
$$= \Sigma_{j=1}^{n} x_j\ \sigma\ (y_j).$$

2) implies 3). Let $x_1 \ldots x_n$; $y_1 \ldots y_n$ be the elements of S given in 2). Define $f_j \in \mathrm{Hom}_R\ (S,R)$ by $f_j(x) = \Sigma_{\sigma \in G}\ \sigma(xy_j)$. Then for all $x \in S$

$$\Sigma_{j=1}^{n} f_j(x)\ x_j = \Sigma_{j=1}^{n} \Sigma_{\sigma \in G}\ \sigma(x)\ \sigma(y_j)\ x_j$$
$$= \Sigma_{\sigma \in G}\ \sigma(x)\ \Sigma_{j=1}^{n} x_j\ \sigma(y_j)$$
$$= x.$$

Thus the set $f_1 \ldots f_n$, $x_1 \ldots x_n$ forms a dual basis for S over R and S is a finitely generated projective R-module.

To prove j is an epimorphism let $h \in \mathrm{Hom}_R\ (S,S)$ and let $w = \Sigma_{\sigma \in G}\ \Sigma_{j=1}^{n} h(x_j)\ \sigma(y_j) u_\sigma$ be an element of $\Delta(S:G)$. Then for any $x \in S$

$$j(w)\ [x] = \Sigma_{\sigma \in G}\ \Sigma_{j=1}^{n} h(x_j)\ \sigma(y_j)\ \sigma(x)$$
$$= \Sigma_{j=1}^{n} h(x_j)\ \Sigma_{\sigma \in G}\ \sigma(y_j x)$$
$$= h\ (\Sigma_{\sigma \in G}\ \Sigma_{j=1}^{n} x_j\ \sigma(y_j)\ \sigma(x))$$
$$= h\ (x).$$

Thus $j(w) = h$ and j is an epimorphism. To prove j is an isomorphism let $w = \Sigma_{\sigma \in G}\ a_\sigma\ u_\sigma \in \Delta\ (S:G)$ and assume $j(w) = 0$. Then

$$0 = \Sigma_{j=1}^{n} \Sigma_{\sigma \in G}\ j(w)\ [x_j]\ \sigma(y_j)\ u_\sigma$$
$$= \Sigma_{\sigma \in G}\ \Sigma_{\tau \in G}\ \Sigma_{j=1}^{n} a_\tau\ \tau(x_j)\ \sigma(y_j)\ u_\sigma$$
$$= \Sigma_{\sigma \in G}\ \Sigma_{\tau \in G}\ a_\tau\ \tau(\Sigma_{j=1}^{n} x_j \tau^{-1}\ \sigma(y_j))\ u_\sigma$$
$$= \Sigma_{\sigma \in G}\ a_\sigma\ u_\sigma = w.$$

3) implies 4). If $x \in S^G$ then x is in the center of $\Delta\ (S:G)$, so $j(x)$ is in the center of $\mathrm{Hom}_R\ (S,S)$. By Proposition 2.4.1 this implies $x \in R$. Thus $S^G = R$.

Now let $t = \Sigma_{\sigma \epsilon G} u_\sigma$ in Δ (S:G). We show that $j(t \cdot S) = \mathrm{Hom}_R(S,R)$. If $a \epsilon S$ and $x \epsilon S$ then $j(t \cdot a)[x] = \Sigma_{\sigma \epsilon G} \sigma(ax) \epsilon R$ so $j(t \cdot S) \subseteq \mathrm{Hom}_R(S,R)$. Suppose $f \epsilon \mathrm{Hom}_R(S,R)$ and $f = j(w)$, $w = \Sigma_{\sigma \epsilon G} a_\sigma u_\sigma$. Then for all $x \epsilon S$, $\Sigma_{\sigma \epsilon G} a_\sigma \sigma(x) \epsilon R$ so for any $\rho \epsilon G$

$$\rho(\Sigma_{\sigma \epsilon G} a_\sigma \sigma(x)) = \Sigma_{\tau \epsilon G} \rho(a_{\rho^{-1}\tau}) \tau(x)$$
$$= \Sigma_{\tau \epsilon G} a_\tau \tau(x)$$

where $\tau = \rho\sigma$. Since j is an isomorphism, $\rho(a_{\rho^{-1}\tau}) = a_\tau$ or $a_\sigma = \sigma(a_1)$ for all $\sigma \epsilon G$. Thus $y = \Sigma_{\sigma \epsilon G} \sigma(a_1) u_\sigma = t \cdot a_1$, and $\mathrm{Hom}_R(S,R) = j(t \cdot S)$. Now we construct the chain of R-module isomorphisms

$$x \otimes y \xrightarrow{f_1} x \otimes t \cdot y \xrightarrow{f_2} x \otimes j(t \cdot y) \xrightarrow{f_3} g_{x \otimes y} \xrightarrow{f_4} \Sigma_{\sigma \epsilon G} x\sigma(y) u_\sigma \xrightarrow{f_5} \Sigma_{\sigma \epsilon G} x\sigma(y) v_\sigma$$

$$S \otimes S \xrightarrow{f_1} S \otimes t \cdot S \xrightarrow{f_2} S \otimes \mathrm{Hom}_R(S,R) \xrightarrow{f_3} \mathrm{Hom}_R(S,S) \xrightarrow{f_4} \Delta(S:G) \xrightarrow{f_5} \nabla(S:G).$$

The isomorphism f_3 is given in Corollary 1.3.4, thus $g_{x \otimes y}[s] = xj(t \cdot y)[s] = j(\Sigma_{\sigma \epsilon G} x\sigma(y) u_\sigma)[s]$. Notice that the composition of these R-module isomorphisms is the R-algebra homomorphism ℓ. Thus ℓ is an isomorphism.

4) implies 2). Let $\ell^{-1}(v_1) = \Sigma_{j=1}^{n} x_j \otimes y_j$, then from the definition of ℓ, $\Sigma_{j=1}^{n} x_j \sigma(y_j) = \delta_{\sigma,1}$.

2) implies 1). Let $x_1 \ldots x_n$, $y_1 \ldots y_n$ be elements of S satisfying ii of 2) and let $tr \epsilon \mathrm{Hom}_R(S,R)$ be defined by $tr(x) = \Sigma_{\sigma \epsilon G} \sigma(x)$. We let $e = \Sigma_{j=1}^{n} x_j \otimes y_j \epsilon S \otimes S$, then $\Sigma_{j=1}^{n} x_j y_j = 1$ and for any $x \epsilon S$

$$(x \otimes 1)e = \Sigma_{j=1}^{n} xx_j \otimes y_j$$
$$= \Sigma_{j=1}^{n} \Sigma_{k=1}^{n} tr(xx_j y_k) x_k \otimes y_j$$
$$= \Sigma_{k=1}^{n} \Sigma_{j=1}^{n} x_k \otimes tr(xy_k x_j) y_j$$
$$= \Sigma_{k=1}^{n} x_k \otimes y_k x$$
$$= (1 \otimes x)e.$$

Therefore by Proposition 2.1.1 S is separable over R. Finally let $\sigma, \tau \in G$, let $0 \neq e = e^2 \in S$, and assume $\sigma(x)e = \tau(x)e$ for all $x \in S$. Then $x\sigma^{-1}(e) = \sigma^{-1}\tau(x)\sigma^{-1}(e)$ for all $x \in S$ so

$$\begin{aligned}\sigma^{-1}(e) &= \Sigma_{j=1}^{n} x_j y_j \sigma^{-1}(e)\\ &= \Sigma_{j=1}^{n} x_j \sigma^{-1}\tau(y_j)\sigma^{-1}(e)\\ &= \delta_{\sigma,\tau}\sigma^{-1}(e).\end{aligned}$$

Since $\sigma^{-1}(e) \neq 0$ we must have $\sigma = \tau$.

2) implies 5). If, for some $\sigma \neq 1$ in G and some maximal ideal M of S, $(1-\sigma)S \subseteq M$, then by 2) we have that $1 = \Sigma_{j=1}^{n} x_j(y_j - \sigma(y_j))$ is in M, a contradiction.

5) implies 2). Let $\sigma \neq 1$ be in G. By 5) the ideal generated by elements of the form $\Sigma_{j=1}^{n} x_j(1-\sigma)y_j$ is all of S. Thus there are elements $x_1 \ldots x_n$, $y_1 \ldots y_n$ in S with $\Sigma_{j=1}^{n} x_j(1-\sigma)y_j = 1$. Let $x_{n+1} = -\Sigma_{j=1}^{n} x_j\sigma(y_j)$ and $y_{n+1} = 1$. Then $\Sigma_{j=1}^{n+1} x_j\sigma(y_j) = \delta_{1,\sigma}$. Next let V and V$'$ be any two subsets of G containing the identity of G and for which there are elements $x_1 \ldots x_n$, $y_1 \ldots y_n$; $x_1' \ldots x_m'$, $y_1' \ldots y_m'$ in S so that for all $\sigma \in V$, $\sigma' \in V'$

$$\Sigma_{j=1}^{n} x_j\sigma(y_j) = \delta_{\sigma,1}$$

$$\Sigma_{k=1}^{m} x_k'\sigma'(y_k') = \delta_{\sigma',1}$$

then for any $\tau \in V \cup V'$

$$\Sigma_{(j,k)=(1,1)}^{(n,m)} x_j x_k' \tau(y_k' y_j) = \delta_{\tau,1}.$$

Since $G = \bigcup_{\sigma\neq 1}\{1,\sigma\}$ the proof is complete.

<u>Definition of a Galois Extension:</u> Let S be an extension of R and let G be a finite group of automorphisms of S, then S is called a <u>Galois extension</u> of R with <u>Galois group</u> G in case one of the five equivalent conditions of Proposition 1.2 is satisfied. Observe that if 0 and 1

are the only idempotents in S then condition 1) ii of Proposition 1.2 is always satisfied so if S is a Galois extension of R then S is a normal separable extension of R. Also recall that in the proof of 3) implies 4) of Proposition 1.2 we showed that if S is a Galois extension of R with group G and $tr = j\ (\Sigma_{\sigma \in G} u_\sigma)$ then tr is a free generator of $Hom_R\ (S,R)$ viewed as a right S-module.

Corollary 1.3: Let S be a Galois extension of R with group G, then

1) There is an element $c \in S$ with $tr(c) = 1$.

2) R is an R-direct summand of S.

3) If T is any commutative R-algebra and G operates on $T \otimes S$ by $\sigma(t \otimes s) = t \otimes \sigma(s)$ for all $\sigma \in G$, $t \in T$, $s \in S$; then $T \otimes S$ is a Galois extension of T with group G.

4) $Rank_R\ (S) = [G:1]$.

Proof: S is finitely generated projective and faithful over R so by Corollary 1.1.9 S is a generator module and there exist $x_1 \ldots x_n$ in S and $f_1 \ldots f_n$ in $Hom_R\ (S,R)$ with $\Sigma_{i=1}^{n} f_i(x_i) = 1$. Since S is a Galois extension of R, tr generates $Hom_R\ (S,R)$ as a right S-module and therefore $f_i = tr(c_i -)$ for some $c_i \in S$. Thus

$1 = \Sigma_{i=1}^{n} tr(c_i x_i) = tr\ (\Sigma_{i=1}^{n} c_i x_i)$ so just let

$c = \Sigma_{i=1}^{n} c_i x_i$.

To prove 2) note that R is a projective R-module and that the sequence $S \xrightarrow{tr} R \to 0$ is exact by 1).

To prove 3) note that we can identify T with $T \otimes 1$ in $T \otimes S$ since R is an R-direct summand of S. If $x_1 \ldots x_n$, $y_1 \ldots y_n$ are in S with $\Sigma_{j=1}^{n} x_j\ \sigma(y_j) = \delta_{\sigma,1}$ then $1 \otimes x_1, \ldots, 1 \otimes x_n$, $1 \otimes y_1, \ldots, 1 \otimes y_n$ are in $T \otimes S$ and $\Sigma_{j=1}^{n} (1 \otimes x_j)\ 1 \otimes \sigma(1 \otimes y_j) = \delta_{\sigma,1}$ in $T \otimes 1 = T$. To complete the verification of 2) of Proposition 1.2 we show $T = (T \otimes S)^G$. Clearly $T \subseteq (T \otimes S)^G$, let $w \in (T \otimes S)^G$ and observe that if $c \in S$ with $tr(c) = 1$ then $\sum_{\sigma \in G} (1 \otimes \sigma)(1 \otimes c) = 1 \otimes 1$. Now let $w \cdot 1 \otimes c = \sum_{i=1}^{n} t_i \otimes s_i$ in $T \otimes S$, then

$$\begin{aligned} w = w \cdot 1 \otimes 1 &= w \sum_{\sigma \in G} 1 \otimes \sigma \, (1 \otimes c) \\ &= \sum_{\sigma \in G} 1 \otimes \sigma(w \cdot 1 \otimes c) \\ &= \sum_{\sigma \in G} 1 \otimes \sigma(\sum_{i=1}^{n} t_i \otimes s_i) \\ &= \sum_{i=1}^{n} t_i \otimes tr(s_i) \\ &= \sum_{i=1}^{n} t_i \, tr(s_i) \otimes 1 \in T \otimes 1. \end{aligned}$$

To verify 4) assume first that R is a local ring. Then both S and $Hom_R(S,S)$ are free R-modules and $Hom_R(S,S) \simeq \Delta(S:G)$ by 3) of Proposition 1.2. Thus

$$\begin{aligned} Rank_R(S)^2 &= Rank_R(Hom_R(S,S)) \\ &= Rank_R(\Delta(S:G)) \\ &= [G:1] \, Rank_R(S). \end{aligned}$$

Thus $Rank_R(S) = [G:1]$ when R is local. If R is an arbitrary commutative ring and $\mathfrak{p}$ is a prime ideal of R then by 3) of this corollary $R_{\mathfrak{p}} \otimes S$ is a Galois extension of $R_{\mathfrak{p}}$ with group G so $[G:1] = Rank_{R_{\mathfrak{p}}}(R_{\mathfrak{p}} \otimes S)$. This proves $[G:1] = Rank_R(S)$.

Corollary 1.4: Let S be an extension of R, let G be a finite group of automorphisms of S with $S^G = R$. Assume that S is a field, then S is a Galois extension of R.

Proof: The verification of 5) of Proposition 1.2 is immediate.

Corollary 1.4 points out the relevance of our general approach to the standard Galois theory for fields. We will now prove a result of a general nature which is a fundamental tool and will be used repeatedly throughout the remainder of this chapter.

Lemma 1.5: Let T be a commutative separable R-algebra and let $h: T \to R$ be an R-algebra homomorphism. Then there exists a unique idempotent $e \in T$ such that

i) $h(t)e = te$ for all $t \in T$

ii) $h(e) = 1$.

Moreover if R has no idempotents other than 0 and 1, if $h_i (i = 1,\ldots,n)$ are distinct algebra homomorphisms from T to R, and if $e_i (i = 1,\ldots,n)$ are the corresponding idempotents then

i) $e_i e_j = e_j \delta_{i,j}$

ii) $h_i(e_j) = \delta_{i,j}$

<u>Proof:</u> Let $\Sigma_{j=1}^{n} x_j \otimes y_j$ be a separability idempotent for T, and let $\mu: T \otimes T \to T$ by $\mu(x \otimes y) = xy$. Given the algebra homomorphism $h: T \to R$ let $e = \Sigma_{j=1}^{n} h(x_j)\, y_j$. Observe that e is the homomorphic image of an idempotent and thus is itself an idempotent. Also for any $t \in T$

$$\begin{aligned} te &= 1 \otimes t\ \mu(\Sigma_{j=1}^{n} h(x_j) \otimes y_j) \\ &= \mu\cdot h \otimes 1\ (\Sigma_{j=1}^{n} x_j \otimes y_j t) \\ &= \mu\cdot h \otimes 1\ (\Sigma_{j=1}^{n} tx_j \otimes y_j) \\ &= \Sigma_{j=1}^{n} h(tx_j) y_j \\ &= h(t)e. \end{aligned}$$

Also since h is an R-homomorphism from T to R we have

$$\begin{aligned} h(e) &= h(\Sigma_{j=1}^{n} h(x_j) y_j) \\ &= \Sigma_{j=1}^{n} h(x_j)\ h(y_j) \\ &= h(1) = 1. \end{aligned}$$

If e' is another idempotent with $te' = h(t)e'$ for all $t \in T$ and $h(e') = 1$ then $ee' = h(e)e' = e' = e'e = h(e')e = e$ which proves the uniqueness statement.

Suppose R has no idempotents other than 0 and 1 and $h_i (j = 1,\ldots,n)$ are algebra homomorphisms from T to R and

$e_i (i = 1,\ldots,n)$ are the corresponding idempotents. Then $h_j(e_i)$ is an idempotent in R and so is either 0 or 1. If $h_j(e_i) = 1$ then for all $t \in T$

$$\begin{aligned} h_j(t) &= h_j(t\cdot 1) = h_j(t)\ h_j(1) \\ &= h_j(t)\ h_j(e_i) = h_j(te_i) \\ &= h_j(h_i(t)e_i) = h_i(t)\ h_j(e_i) \\ &= h_i(t) \end{aligned}$$

and $h_j = h_i$ so $i = j$. Thus $h_i(e_j) = \delta_{i,j}$. Finally, $e_i e_j = h_j(e_i)e_j = \delta_{i,j}e_j$ which completes the proof.

Some immediate consequences of Lemma 1.5 are contained in

Corollary 1.6: Let S be a finitely generated commutative separable R-algebra and let T be a commutative R-algebra with no idempotents other than 0 and 1, then there are only finitely many R-algebra homomorphisms from S to T. If S is also a projective R-algebra and Rank_R (S) is defined then there are at most Rank_R (S) homomorphisms from S to T. If S is a Galois extension of R with group G and if g and f are two algebra homomorphisms from S to T then there is a unique σ in G with $g = f\sigma$.

Proof: Since S is a finitely generated separable R-algebra by Corollary 2.1.7 $T \otimes S$ is a finitely generated separable T-algebra. Let $\sigma_1,\ldots,\sigma_m$ be distinct algebra homomorphisms from S to T, then the homomorphisms $h_1,\ldots,h_m$ defined by $h_i(x \otimes y) = x\sigma_i(y)$ are distinct T-algebra homomorphisms from $T \otimes S$ to T. By Lemma 1.5 there are orthogonal idempotents $e_1,\ldots,e_m$ in $T \otimes S$ with $we_i = h_i(w)e_i$ for all $w \in T \otimes S$ and $h_i(e_j) = \delta_{i,j}$. Thus $(T \otimes S)e_i = (T \otimes 1)e_i$, and $(x \otimes 1)e_i = 0$ implies $(x \otimes 1)\ h_i(e_i) = x = 0$. Therefore we can identify $(T \otimes S)e_i$ with T. Since the e_i are orthogonal

$$\begin{aligned} T \otimes S &\simeq \underbrace{(T \otimes S)\ e_1 \oplus\ldots\oplus (T \otimes S)\ e_m}_{\text{-m-times-}} \oplus (T \otimes S)(1-e_1-\ldots-e_m) \\ &\simeq T \oplus\ldots\oplus T \oplus (T \otimes S)(1-e_1-\ldots-e_m). \end{aligned}$$

Let M be any maximal ideal of T. Then from the decomposition above $T/M \otimes S$ must be a vector space whose dimension is greater than or equal to m. Thus if S is generated by k-elements over R then $m \leq k$. Moreover if S is projective and has rank n over R then $m \leq n$. If S is a Galois extension of R with group G then $m \leq [G:1]$. If f is an R-homomorphism from S to T and $\sigma \in G$ with $f\sigma = \sigma$ then $f(\sigma(x) - x) = 0$ for all $x \in S$ so Proposition 1.2 implies $\sigma = 1$. Therefore the set $\{f\sigma | \sigma \in G\}$ contains $[G:1]$ distinct elements, and by the pidgeon hole principle if g is another R-homomorphism from S to T there is a unique $\sigma \in G$ with $g = f\sigma$.

<u>Corollary 1.7</u>: Let S be a Galois extension of R with group G and assume 0 and 1 are the only idempotents in S, then any R-homomorphism from S to S is an automorphism in G.

<u>Proof</u>: If f is an R-homomorphism from S to S and $\sigma \in G$ then by Corollary 1.6 there is a $\tau \in G$ with $f = \sigma\tau$, thus $f \in G$.

<u>Proof of Theorem 1.1</u>: It is now possible to completely prove Theorem 1.1. With the hypothesis and notation of Theorem 1.1 we have by Corollary 1.6 that $[G:1]$ is finite so by Proposition 1.2(1) S is a Galois extension of R with group G. By Corollary 1.3 we have $[G:1] = \text{Rank}_R\ (S)$.

Next we show the correspondence $H \to S^H$ is one to one from the subgroups of G to the subrings of S containing R and separable over R. By Proposition 1.2 (2) there exist $x_1, \ldots, x_n, y_1, \ldots, y_n$ in S with $\sum_{j=1}^{n} x_j\sigma(y_j) = \delta_{1,\sigma}$ for all $\sigma \in G$. Using this condition we have that S is a Galois extension of S^H so by Corollary 1.7 if K is a subgroup of G with $H \neq K$ then $S^H \neq S^K$. Let $tr \in \text{Hom}_{S^H}(S, S^H)$ be the trace map given by $tr = \Sigma_{\sigma \in H}\sigma$. By Corollary 1.3 there is an element $c \in S$ with $tr(c) = 1$. Let $a_j = tr(x_j)$, $b_j = tr(y_jc)$ and let $e = \sum_{j=1}^{n} a_j \otimes b_j \in S^H \otimes S^H$. Then

$$\sum_{j=1}^{n} a_jb_j = \sum_{j=1}^{n} \Sigma_{\sigma \in H} \Sigma_{\tau \in H}\ \sigma(x_j)\tau(y_jc) = \sum_{\sigma\tau}[\sum_j x_j\sigma^{-1}\tau(y_j)]\sigma^{-1}\tau(c) = 1.$$

Also for any $x \in S^H$, since $\Sigma_{j=1}^{n} xx_j \otimes y_j = \Sigma_{j=1}^{n} x_j \otimes y_j x$ we have

$$\begin{aligned}(x \otimes 1)e &= \Sigma_{j=1}^{n} xa_j \otimes b_j \\ &= \Sigma_{j=1}^{n} \Sigma_{\sigma\in H} \Sigma_{\tau\in H} x\sigma(x_j) \otimes \tau(y_j c) \\ &= \Sigma_{\sigma\in H} \Sigma_{\tau\in H} \sigma \otimes \tau \; (\Sigma_{j=1}^{n} xx_j \otimes y_j \cdot 1 \otimes c) \\ &= \Sigma_{j=1}^{n} (\Sigma_{\sigma\in H} \sigma(x_j)) \otimes (\Sigma_{\tau\in H} \tau(y_j c)) \; 1 \otimes x \\ &= (1 \otimes x) \Sigma_{j=1}^{n} a_j \otimes b_j \\ &= (1 \otimes x)e .\end{aligned}$$

Thus by Proposition 2.1.1 S^H is separable over R and the correspondence $H \to S^H$ is one to one from the subgroups of G to the subrings of S containing R and separable over R. Observe S is a Galois extension of S^H with group H so Corollary 1.3 implies S is a finitely generated extension of S^H. If H is a normal subgroup of G, then G/H acts as a group of automorphisms of S^H by $\sigma H\,(x) = \sigma(x)$ for all $\sigma H \in G/H$ and $x \in S^H$. Observe that $(S^H)^{G/H} = R$ so S^H is a normal extension of R.

Now let T be a subring of S containing R and separable over R. Let $H = \{\sigma\in G \,|\, \sigma(x) = x \text{ for all } x\in T\}$, then H is a subgroup G and $T \subseteq S^H$. Let $\sigma_1,\ldots,\sigma_n$ be a full set of distinct left coset representatives of H in G with $\sigma_1 = 1$. Let h_i be the distinct S-algebra homomorphisms from $S \otimes T$ to S defined by $h_i\,(x \otimes y) = x\sigma_i(y)\,(i = 1,\ldots,n)$. By Corollary 2.1.7, $S \otimes T$ is a separable S-algebra and S has no idempotents other than 0 and 1 so by Lemma 1.5 there is an idempotent e_1 in $S \otimes T$ with $h_1(w)e_1 = we_1$ and $h_i(e_1) = \delta_{i,1}$ for all $w \in S \otimes T$, $i = 1,\ldots,n$. Let $e_1 = \Sigma_{j=1}^{m} s_j \otimes t_j$, then $h_i(e_1) = \Sigma_{j=1}^{n} s_j\, \sigma_i(t_j) = \delta_{1,i}$ so

$$\Sigma_{j=1}^{n} s_j\, \sigma(t_j) = \begin{cases} 1 & \sigma \in H \\ 0 & \sigma \notin H . \end{cases}$$

Applying σ^{-1} to both sides of this equation and relabeling we have

$$\Sigma_{j=1}^{n} \tau(s_j)\, t_j = \begin{cases} 1 & \tau \in H \\ 0 & \tau \notin H . \end{cases}$$

Let $c \in S$ with $\Sigma_{\sigma \in H} \sigma(c) = 1$ and let $a \in S^H$, then

$$a = \Sigma_{\sigma \in H} \sigma(ca)$$
$$= \Sigma_{\sigma \in G} \Sigma_{j=1}^{n} \sigma(s_j ca) t_j$$
$$= \Sigma_{j=1}^{n} (\Sigma_{\sigma \in G} \sigma(s_j ca) t_j .$$

Since $\Sigma_{\sigma \in G} \sigma(s_j ca) \in R$ and $t_j \in T$ we have that $a \in T$, and $T = S^H$. Finally let T be a normal separable extension of R in S and let $T = S^H$. Observe that $\sigma H \sigma^{-1} = H$ for all $\sigma \in G$ if and only if $\sigma(T) = T$ for all $\sigma \in G$. There are Rank_R (T) automorphisms of T since T is a finite normal separable extension of R, therefore by Corollary 1.6 any homomorphism from T to S is an automorphism of T. In particular $\sigma(T) = T$ for all $\sigma \in G$ which completes the proof.

§ 2. The Imbedding Theorem

In this section we show that any finite, projective, separable, extension with no idempotents other than 0 and 1 can be imbedded in a normal separable extension whose only idempotents are 0 and 1. Along the way to a proof of this result some interesting properties of separable projective extensions will be exhibited.

To begin let A be any R-algebra and let M be a left A-module which is finitely generated and projective as an R-module. Let $x_1, \ldots, x_m \in M$ and $f_1, \ldots, f_m \in \text{Hom}_R (M,R)$ be an R-dual basis for M. Define $t_M \in \text{Hom}_R (A,R)$ by

$$t_M(x) = \Sigma_{i=1}^{m} f_i(xx_i), \ x \in A .$$

We call t_M the <u>trace</u> from A to R afforded by M. An easy computation (Exercise 3.1) shows t_M is independent of the choice of a dual basis for M. If $M = M_1 \oplus M_2$ as A-modules it is easy to see that $t_M = t_{M_1} + t_{M_2}$ and if $M = A = R$ then $t_M(x) = x$ for all x. The next theorem points out the importance of the trace in the study of separable algebras.

Theorem 2.1: The extension S of R is finite separable and projective over R if and only if there is an element $t \in \mathrm{Hom}_R(S,R)$ and elements $x_1,\ldots,x_n, y_1,\ldots,y_n$ in S with

$$\text{i. } \sum_{j=1}^{n} x_j y_j = 1$$

$$\text{ii. } \sum_{j=1}^{n} x_j t(y_j x) = x \text{ for all } x \in S.$$

Moreover the map t is always the trace map from S to R.

Proof: Let S be a finite, projective, separable extension of R, and let $a_1,\ldots,a_m, f_1,\ldots,f_m$ be an R-dual basis of the R-module S. Then the trace map from S to R is given by $t_S(x) = \sum_{j=1}^{m} f_j(xa_j)$ for all $x \in S$. Since S is a finite, projective, separable extension of R, $S \otimes S$ is a finite projective separable extension of $S \otimes 1$ with a dual basis $1 \otimes a_1,\ldots,1 \otimes a_m, 1 \otimes f_1,\ldots,1 \otimes f_m$ and trace map $t = 1 \otimes t_S$ mapping $S \otimes S$ to $S \otimes 1$. Since S is separable over R we have by Proposition 2.1.1 that $S \otimes S \simeq J \oplus (S \otimes S)e$ as $S \otimes S$-modules where e is a separability idempotent for S. Since $(1 \otimes x - x \otimes 1)e = 0$ for all $x \in S$ and since $(S \otimes S)e \simeq S$ under the homomorphism μ of Proposition 2.1.1 we can write $S \otimes S \simeq J \oplus (S \otimes 1)e \simeq J \oplus S$. This direct sum decomposition allows us to write $t = t_J + t_e$ where t_J is t restricted to J and t_e is t restricted to $(S \otimes 1)e$. A consequence of our identification of $(S \otimes 1)e$ with S is that $t_e[(x \otimes 1)e] = x$ for all $x \in S$. Now let $x \in S$ and let $e = \sum_{j=1}^{n} x_j \otimes y_j$, then

$$\begin{aligned} x &= t(x \otimes 1 \cdot e) \\ &= t(1 \otimes x \cdot e) \\ &= 1 \otimes t_S\Big(\sum_{j=1}^{n} x_j \otimes y_j x\Big) \\ &= \sum_{j=1}^{n} x_j \otimes t_S(y_j x) \\ &= \sum_{j=1}^{n} x_j t_S(y_j x) \otimes 1 \\ &= \sum_{j=1}^{n} x_j t_S(y_j x). \end{aligned}$$

Since $e = \Sigma_{j=1}^{n} x_j \otimes y_j$ is a separability idempotent for S, $\Sigma_{j=1}^{n} x_j y_j = 1$ and this proves the direct part of the equivalence.

To prove the converse observe that if $t \in \mathrm{Hom}_R(S,R)$ and $x_1,\ldots,x_n$, $y_1,\ldots,y_n$ in S are given with $\Sigma_{j=1}^{n} x_j y_j = 1$ and $x = \Sigma_{j=1}^{n} x_j\, t(y_j x)$ for all $x \in S$ then the set $x_1,\ldots,x_n$, $t(y_1-),\ldots,t(y_n-)$ forms a dual basis for S over R, so S is a finitely generated, projective R-module. Let $e = \Sigma_{j=1}^{n} x_j \otimes y_j \in S \otimes S$, then $\Sigma_{j=1}^{n} x_j y_j = 1$ and for any $x \in S$

$$
\begin{aligned}
1 \otimes x \cdot e &= \Sigma_{j=1}^{n} x_j \otimes y_j x \\
&= \Sigma_{j=1}^{n} \Sigma_{i=1}^{n} x_j \otimes x_i t(y_i y_j x) \\
&= \Sigma_{i=1}^{n} \Sigma_{j=1}^{n} x_j t(y_j y_i x) \otimes x_i \\
&= \Sigma_{i=1}^{n} x y_i \otimes x_i \\
&= x \otimes 1 \cdot e
\end{aligned}
$$

where $\Sigma_{j=1}^{n} x_j \otimes y_j = \Sigma_{i=1}^{n} y_i \otimes x_i$ by letting $x = 1$ in the equations above. Thus by Proposition 2.1.1 S is also separable over R.

The homomorphism t must be the trace map from S to R since for any $x \in S$ the trace map at x is given by $\Sigma_{j=1}^{n} t(y_j x x_j) = t(x)$.

<u>Corollary 2.2:</u> The finite projective extension S of R is separable if and only if the trace map t from S to R is a free right S-module generator of $\mathrm{Hom}_R(S,R)$.

Proof: If S is a finite projective separable extension of R, if t is the trace map from S to R, and if $x_1,\ldots,x_n$, $y_1,\ldots,y_n$ are the elements given in Theorem 2.1 then for any $f \in \mathrm{Hom}_R(S,R)$ and any $x \in S$

$$
\begin{aligned}
f(x) &= \Sigma_{j=1}^{n} f(x_j t(y_j x)) \\
&= \Sigma_{j=1}^{n} t(y_j x)\, f(x_j) \\
&= t(\Sigma_{j=1}^{n} y_j x\, f(x_j)).
\end{aligned}
$$

If $a = \Sigma_{j=1}^{n} f(x_j)y_j$ then $f(x) = t(ax)$ for all $x \in S$. If $t(a-) = 0$ in $\mathrm{Hom}_R(S,R)$ then

$$0 = \Sigma_{j=1}^{n} x_j t(y_j a) = a$$

so t is a free right S-module generator of $\mathrm{Hom}_R(S,R)$. Conversely if $x_1,\dots,x_m$, $f_1,\dots,f_m$ forms a dual basis of S over R and we can find $y_j \in S$ with $f_j = t(y_j -)$ then for any $x \in S$

$$x = \Sigma_{j=1}^{m} f_j(x)\ x_j = \Sigma_{j=1}^{m} x_j t(y_j x)$$

for any $z \in S$

$$\begin{aligned} t[(1-\Sigma_{j=1}^{m} x_j y_j)z] &= t(z) - t(\Sigma_{j=1}^{m} x_j y_j z) \\ &= \Sigma_{j=1}^{m} f_j(zx_j) - t(\Sigma_{j=1}^{m} x_j y_j z) \\ &= t(\Sigma_{j=1}^{m} zx_j y_j) - t(\Sigma_{j=1}^{m} x_j y_j z) \\ &= 0. \end{aligned}$$

Since t is a free generator of $\mathrm{Hom}_R(S,R)$ we conclude that $\Sigma_{j=1}^{m} x_j y_j = 1$.

In the special case that S is a Galois extension of R the trace map t of Corollary 2.2 is given by $t(x) = \Sigma_{\sigma\in G}\ \sigma(x)$ for all $x \in S$.

Corollary 2.3: If S is a finite projective separable extension of R and t is the trace map from S to R then there is an element $c \in S$ with $t(c) = 1$. Moreover $R\cdot 1$ is an R-direct summand of S.

Proof: The hypotheses on S imply S is a generator module over R so there are elements $f_1,\dots,f_n$ in $\mathrm{Hom}_R(S,R)$ and $x_1,\dots,x_n$ in S with $1 = \Sigma_{j=1}^{n} f_j(x_j)$. By Corollary 2.2 there is for each j an element $a_j \in S$ with $f_j(x_j) = t(a_j x_j)$, so $1 = t(\Sigma_{j=1}^{n} a_j x_j)$. Thus the sequence $S \xrightarrow{t} R \to 0$ is exact and must split since R is R-projective.

We next have the important

Theorem 2.4: Let S be an extension of R and assume T is an extension of R contained in S, then any two of the following statements imply the third.

1) T is a finite separable projective extension of R.

2) S is a finite separable projective extension of T.

3) S is a finite separable projective extension of R.

Proof: 1) + 2) implies 3). Let $t_{S/T}$ and $t_{T/R}$ be the trace maps from S to T and T to R respectively. Let $x_1,\dots,x_n, y_1,\dots,y_n \in S$ and $a_1,\dots,a_m, b_1,\dots,b_m \in T$ with

$$1 = \Sigma_{j=1}^{n} x_j y_j = \Sigma_{j=1}^{m} a_j b_j$$

and

$$x = \Sigma_{j=1}^{n} x_j \, t_{S/T}(y_j x) \quad \text{for all } x \in S$$

$$y = \Sigma_{j=1}^{m} a_j \, t_{T/R}(b_j y) \quad \text{for all } y \in T .$$

Let $t = t_{T/R} \cdot t_{S/T}$, then $t \in \mathrm{Hom}_R(S,R)$. Let $w_{ij} = x_i a_j$ and $z_{ij} = y_i b_j$, $i = 1,\dots,n$, $j = 1,\dots,m$. Then $w_{ij}, z_{ij} \in S$ and

$$\Sigma_{i=1}^{n} \Sigma_{j=1}^{m} w_{ij} z_{ij} = 1$$

$$\Sigma_{i=1}^{n} \Sigma_{j=1}^{m} w_{ij} \, t(z_{ij} x) = x \text{ for all } x \in S.$$

By Theorem 2.1 S is a finite separable projective extension of R.

1) + 3) implies 2). S is a finite extension of R so S is a finite extension of T, S is projective over R and T is separable over R so by Proposition 2.2.3 S is projective over T. Also S is separable over R so by Proposition 2.1.12 S is separable over T.

2) + 3) implies 1). By Corollary 2.3 T is a T-direct summand of S so T is an R-direct summand of S so T is a finite projective extension of R. Now S is projective over T so $S \otimes S$ is projective over $T \otimes T$. But $S \otimes S \simeq J \oplus (S \otimes S)e$ since S is separable over R and we saw in the proof of Theorem 2.1 that $(S \otimes S)e \simeq S$ so $S \otimes S \simeq S \oplus J$. This is an isomorphism as $S \otimes S$-modules and T is a direct summand of S so $T \otimes T \subseteq S \otimes S$ and this is a $T \otimes T$-isomorphism. Thus S is projective over $T \otimes T$. Now T is a $T \otimes T$-direct summand of S so T is a

projective $T \otimes T$-module which by Proposition 2.1.1 implies T is separable over R, completing the proof.

Proposition 2.5: Let S be a finite projective separable extension of R and let T be a separable extension of R contained in S, then T is also a finite projective extension of R.

Proof: S is a finite projective extension of R and T is a separable extension of R so by Proposition 2.2.3 S is a finite projective extension of T. By Proposition 2.1.12 we know S is a separable extension of T so by Corollary 2.3 T is a T-direct summand of S. Therefore T is finitely generated and projective over R.

Corollary 2.6: Let S be a separable R-algebra, let T be a finite projective separable extension of R, and let f be an R-algebra homomorphism from S to T. Then the kernel of f is idempotent generated, thus if S has no idempotents other than 0 and 1 then f is a monomorphism.

Proof: We have the exact sequence

$$0 \to \text{kernel } (f) \to S \overset{f}{\to} f(S) \to 0.$$

By Proposition 2.1.11 f(S) is separable over R so by Proposition 2.5 f(S) is also finitely generated and projective over R. Since S is a separable R-algebra, f(S) is projective over S by Proposition 2.2.3. Therefore our exact sequence splits as S-modules which means kernel (f) is an S-ideal direct summand of S and so is idempotent generated. The last statement of the corollary is clear.

Lemma 2.7: Let S be a finite projective separable extension of R and let T be an extension of R whose only idempotents are 0 and 1. Then there exist Rank_R (S) distinct R-algebra homomorphisms from S to T if and only if $S \otimes T \simeq T \oplus \ldots \oplus T$ (Rank_R (S)-times) as T-algebras.

Proof: Let $\sigma_1,\ldots,\sigma_m$ be distinct R-algebra homomorphisms from S to T where $m = \text{Rank}_R(S)$. Let $\pi_i\colon S \otimes T \to T$ be defined by $\pi_i(s \otimes t) = \sigma_i(s)t$. The π_i are m distinct T-algebra homomorphisms from $S \otimes T$ to T. By Lemma 1.5 there are orthogonal idempotents $e_1,\ldots,e_m$ in $S \otimes T$ with $(S \otimes T)e_i \simeq T$ and $S \otimes T = (S \otimes T)e_1 \oplus \ldots \oplus (S \otimes T)e_m \oplus (S \otimes T)(1-e_1-e_2-\ldots-e_m)$. By a count of ranks, $1-e_1-e_2-\ldots-e_m = 0$.

To prove the converse let $\pi_i\colon S \otimes T \to T$ be the T-algebra projection onto the ith component of the given direct decomposition and let σ_i be the restriction of π_i to $S \otimes 1$. Then one quickly checks that $\sigma_1,\ldots,\sigma_m$ are the required R-algebra homomorphisms.

The imbedding theorem will be a direct consequence of the next lemma.

Lemma 2.8: Let S be a finite projective separable extension of R and assume R has no idempotents other than 0 and 1. Then there exists a finite projective separable extension T of R such that T has no idempotents other than 0 and 1 and

$$T \otimes S = T \oplus \ldots \oplus T$$

where there are $\text{Rank}_R(S)$ summands on the right.

Proof: Since R has no idempotents other than 0 and 1 and since $\text{Rank}_R(S)$ is finite we can write $S = Se_1 \oplus \ldots \oplus Se_m$ where Se_i is a finite projective extension of R with no idempotents other than 0 and e_i. Each Se_i is separable over R by Proposition 2.1.11. Let $m_R(S) = \max\{\text{Rank}_R(Se_i) \mid i = 1,\ldots,m\}$ and let $n_R(S)$ be the number of summands Se_j with $\text{Rank}_R(Se_j) = m_R(S)$. If $m_R(S) = n_R(S) = 1$ then $S = R$ and the result is trivial. Now assume inductively that the lemma is true for an extension B of A satisfying the hypothesis of the lemma and for which $m_A(B) < m_R(S)$ or $m_A(B) = m_R(S)$ and $n_A(B) < n_R(S)$.

Let A denote one of the summands Se_j for which $\text{Rank}_R(Se_j) = m_R(S)$. Then consider the A-algebra $B = A \otimes S$. Since $\text{Rank}_R(Se_j) =$

$\text{Rank}_A\ (A \otimes Se_j)$ we see that $m_A\ (B) \leq m_R\ (S)$. If $m_A\ (B) = m_R\ (S)$ then $n_A\ (B) < n_R\ (S)$ since the component $A \otimes A = Se_j \otimes Se$ has a proper idempotent, the separability idempotent for Se_j, otherwise $m_A\ (B) < m_R\ (S)$. In either case by our inductive hypothesis there is a finite projective separable extension T of A so that T has no idempotents other than 0 and 1 with $T \otimes_A A \otimes S = T \otimes S = T \oplus \ldots \oplus T$. By Theorem 2.4 T is a finite separable projective extension of R. The statement about ranks in the lemma follows since $\text{Rank}_R\ (S) = \text{Rank}_T\ (T \otimes S)$ is the number of summands in $T \otimes S = T \oplus \ldots \oplus T$.

We now come to the main result of this section.

Theorem 2.9: Let S be a finite projective separable extension of R with 0 and 1 the only idempotents in S. Then there is a finite normal separable extension N of R with 0 and 1 the only idempotents of N and with S an extension of R in N.

Proof: By Lemma 2.8 there is a finite separable projective extension T of R with T having no idempotents other than 0 and 1 and with $T \otimes S \simeq T \oplus \ldots \oplus T$ where there are $n = \text{Rank}_R\ (S)$ summands. By Corollary 2.3 R is an R-direct summand of both S and T so we can identify S with $1 \otimes S$ and T with $T \otimes 1$ in $T \otimes S$. If π_i is the T-algebra projection onto the ith coordinate in the decomposition of $T \otimes S$ then π_i restricted to $1 \otimes S$ is an R-algebra homomorphism from S to T which we denote σ_i. If $\sigma_i = \sigma_j$ on S then $\pi_i = \pi_j$ on $T \otimes S$ and by Corollary 2.6 each σ_i is a monomorphism so $\sigma_1, \ldots, \sigma_n$ are all the distinct monomorphisms from S to T by Corollary 1.6. Let $N = \sigma_1(S) \cdots \sigma_n(S)$, N is a homomorphic image of $\sigma_1(S) \otimes \ldots \otimes \sigma_n(S)$ so by Proposition 2.1.6 and Proposition 2.1.11 N is a separable R-subalgebra of T. By Proposition 2.5 N is a finite projective separable extension of R. To show N is a normal extension of R we find for each $x \in N-R$ an R-automorphism σ of N with $\sigma(x) \neq x$.

There are $n = \mathrm{Rank}_R(S)$ N-algebra homomorphisms h_i from $N \otimes S$ to N defined by $h_i(x \otimes y) = x\sigma_i(y)$. By Lemma 2.7

$$N \otimes S = \overbrace{N \oplus \ldots \oplus N}^{\text{-n-times-}}.$$

Now $N \otimes N$ is a homomorphic image of $N \otimes (\overbrace{S \otimes \ldots \otimes S}^{\text{n-times}}) = \overbrace{N \oplus \ldots \oplus N}^{n^n\text{-times}}$ under the homomorphism $(x \otimes s_1 \otimes \ldots \otimes s_n) \to x \otimes \prod_{i=1}^{n} \sigma_i(s_i)$. By Corollary 2.6 the kernel of this homomorphism is generated by an idempotent so $N \otimes N$ is a direct summand of a direct sum of copies of N. Since N has no idempotents other than 0 and 1 we have $N \otimes N = N \oplus \ldots \oplus N$ where there are $m = \mathrm{Rank}_R(N)$ copies of N in the decomposition. Let π_i be the N-algebra projection of $N \otimes N$ onto the ith coordinate in the direct sum decomposition of $N \otimes N$. Define an R-algebra homomorphism τ_i from N to N by letting τ_i be the restriction of π_i to $1 \otimes N$. By Corollary 2.6 τ_i is a monomorphism, and $\tau_i(N)$ is a finite projective separable extension of R in N with $\mathrm{Rank}_R(N) = \mathrm{Rank}_R(\tau_i(N))$. Therefore by Theorem 2.4 N is a finite projective separable extension of $\tau_i(N)$ of rank = 1 and so by Corollary 2.3 $\tau_i(N) = N$. We have constructed a set $\tau_1, \ldots, \tau_m$ of automorphisms of N. By Corollary 1.6 there are no more than m distinct R-automorphisms of N and if $\tau_i = \tau_j$ then $\pi_i = \pi_j$ so $\tau_1, \ldots, \tau_m$ are all the R-automorphisms of N and form a group G. Now N is a separable extension of N^G by Proposition 2.1.12 so N is a normal separable extension of N^G and by Theorem 1.1 $\mathrm{Rank}_{N^G}(N) = [G:1] = \mathrm{Rank}_R(N)$. Clearly $R \subseteq N^G$ and by Theorem 2.4 N^G is a finite projective separable extension of R with $\mathrm{Rank}_R(N^G) = 1$. Therefore by Corollary 2.3 $R = N^G$ and the proof is complete.

3. The Separable Closure and Infinite Galois Theory

Let R be a commutative ring with no idempotents other than 0 and 1. In this section a certain extension is associated with R. This extension will be called the separable closure of R and is nothing more than the algebraic closure of R when R is a perfect field. In order to carry out necessary constructions we will be required to employ some basic facts from set theory and point set topology which can be found in J. L. Kelley's book "General Topology." Throughout this section R and Ω will be commutative rings whose only idempotents are 0 and 1, Ω will be an extension of R.

We begin with some preliminaries. A set D is called a <u>directed set</u> in case there is a reflexive, antisymmetric, transitive partial order "$\leq$" on D such that for any two elements a, b $\in$ D there is an element c $\in$ D with $a \leq c$ and $b \leq c$. A set D is called a <u>linearly ordered</u> set in case there is a reflexive, antisymmetric, transitive order "$\leq$" on D such that for any two elements a, b $\in$ D either $a \leq b$ or $b \leq a$. Notice that any linearly ordered set is a directed set.

Let D be a directed set. Suppose that for each i $\in$ D there is a ring A_i and for each pair i, j $\in$ D with $i \leq j$ there is a ring monomorphism $f_{i,j}:A_i \to A_j$ such that $f_{i,i}$ is the identity on A_i for all i $\in$ D and $f_{j,k}\, f_{i,j} = f_{i,k}$ for all i,j,k $\in$ D with $i \leq j \leq k$. The collection of rings A_i and maps $f_{i,j}$ is called a <u>directed system</u> of rings.

<u>Lemma 3.1</u>: Let D be a directed set and let $\{A_i, f_{i,j}\}$ be a directed system of rings, then there exists a ring $\overline{A}$ with subrings $\overline{A}_i$ and ring isomorphisms $h_i:A_i \to \overline{A}_i$ for all i $\in$ D such that

1) $\overline{A}_i \subseteq \overline{A}_j$ whenever $i \leq j$;

2) $$\begin{array}{ccc} A_i & \xrightarrow{f_{i,j}} & A_j \\ h_i \downarrow & & \downarrow h_j \\ \overline{A}_i & \subseteq & \overline{A}_j \end{array}$$

commutes whenever $i \leq j$;

3) $\bigcup_{i \in D} \overline{A}_i = \overline{A}$.

<u>Proof</u>: By replacing the A_i, if necessary, by isomorphic rings with different underlying sets we may assume $A_i \cap A_j = \emptyset$ when $i \neq j$.

Let $B = \bigcup_{i \in D} A_i$. For a, b ϵ B say $a \sim b$ in case whenever $a \in A_i$, $b \in A_j$ and $i \leq k$, $j \leq k$ then $f_{i,k}(a) = f_{j,k}(b)$. Observe that "$\sim$" is an equivalence relation on B. Let $\bar{A}$ be the set of equivalence classes and for $a \in B$ let $\bar{a}$ be the class containing a. For $a \in A_i$ and $b \in A_j$ define

$$\bar{a} + \bar{b} = \overline{f_{i,k}(a) + f_{j,k}(b)}$$

$$\bar{a} \cdot \bar{b} = \overline{f_{i,k}(a) \cdot f_{j,k}(b)}$$

where $i \leq k$ and $j \leq k$. One can check that these operations are well defined and turn $\bar{A}$ into a ring. For $a \in A_i$ let $h_i(a) = \bar{a}$ and let $\bar{A}_i = h_i(A_i)$, then the conclusion of the lemma is easy to check.

If $\{A_i, f_{i,j}\}$ is a directed system of rings as in Lemma 3.1 the ring $\bar{A}$ constructed in Lemma 3.1 is called the <u>direct limit</u> of the directed system and is denoted $\varinjlim_{i \in D} A_i$.

Again let D be a directed set and let $\{X_i \mid i \in D\}$ be a collection of topological spaces. Let $\prod_{i \in D} X_i$ be the direct product of this collection of topological spaces with the usual product topology. If $\bar{x} \in \prod_{i \in D} X_i$ we let x_k be the image of the projection π_k of $\bar{x}$ on X_k along $\prod_{i \neq k} X_k$. Assume that for all i, j ϵ D with $i \leq j$ there are continuous functions $f_{j,i}: X_j \to X_i$ such that $f_{i,i}$ is the identity on X_i for all $i \in D$ and $f_{ji} f_{kj} = f_{ki}$ for all $i,j,k \in D$ with $i \leq j \leq k$. The collection $\{X_i, f_{i,j}\}$ is called an <u>inverse limit</u> system and

$$\varprojlim_{i \in D} X_i = \{\bar{x} \in \prod_{i \in D} X_i \mid f_{ji}(x_j) = x_i \text{ for all } i \leq j \text{ in } D\}$$

is defined to be the <u>inverse limit</u> of the inverse limit system.

<u>Lemma 3.2</u>: Let D be a directed set and let $\{X_i, f_{i,j}\}$ be an inverse limit system of non-void compact Hausdorff topological spaces, then $\varprojlim_{i \in D} X_i$ is a non-empty compact Hausdorff space.

<u>Proof</u>: For each $j \in D$ let

$$W(j) = \{\bar{x} \in \prod_{i \in D} X_i \mid f_{j,i}(x_j) = x_i \text{ for all } i \leq j\}.$$

Pick $\bar{y} \in \prod_{i \in D} X_i$ but $\bar{y} \notin W(j)$, then there is an $i \in D$ with $i \leq j$ and $f_{j,i}(y_j) \neq y_i$. Since X_i is Hausdorff we can find open sets U_i and V_i in X_i with $U_i \cap V_i = \emptyset$ and with $y_i \in U_i$, $f_{j,i}(y_j) \in V_i$. Let $V_j = f_{j,i}^{-1}(V_i)$, then V_j is an open subset of X_j with $f_{j,i}(V_j) \cap U_i = \emptyset$. Let

$$U = \{\bar{x} \in \prod_{k \in D} X_k \mid x_i \in U_i,\ x_j \in V_j\}$$

then U is an open subset of $\prod_{k \in D} X_k$ and $U \cap W(j) = \emptyset$. Therefore $W(j)$ is a closed subset of $\prod_{k \in D} X_k$. Also observe that

$$W(j) = \bigcup_{x_j \in X_j} \Big(\bigcap_{i \leq j} \pi_i^{-1}(f_{j,i}(x_j))\Big) \neq \emptyset.$$

The family $\{W(j) \mid j \in D\}$ has the finite intersection property since D is a directed set, and $\prod_{i \in D} X_i$ is a compact Hausdorff space by the Tychonoff product theorem so $\bigcap_{j \in D} W(j) \neq \emptyset$. Observe that $\bigcap_{j \in D} W(j) = \varprojlim_{i \in D} X_i$ which proves the lemma.

We have shown a little more than the lemma asserts, namely the $\varprojlim_{i \in D} X_i$ is a non-empty compact Hausdorff space in the topology inherited from $\prod_{i \in D} X_i$. This observation is necessary to prove Theorem 3.6.

<u>Definition of the Separable Closure</u>: Let Ω be an extension of R and assume 0 and 1 are the only idempotents in Ω. We call Ω <u>locally separable</u> in case any finite subset of Ω is contained in a finite separable projective extension of R in Ω. We call Ω <u>separably closed</u> in case there are no finite projective separable extensions of Ω except direct sums of copies of Ω. We call Ω the <u>separable closure</u> of R in case Ω is a locally separable separably closed extension of R.

The principal theorem of this section is

Theorem 3.3: A separable closure Ω of R exists and is unique up to isomorphism. Moreover if S is a finite projective separable extension of R there are $\text{Rank}_R(S)$ algebra homomorphisms from S to Ω. If S has no idempotents other than 0 and 1 the homomorphisms are monomorphisms.

Proof: Suppose that a separable closure does not exist. Then for each non-limit ordinal α we can construct a finite separable projective extension Ω_α of $\Omega_{\alpha-1}$ with 0 and 1 the only idempotents in Ω_α, with $\Omega_\alpha \neq \Omega_{\alpha-1}$, and with $\Omega_{\alpha-1}$ a locally separable extension of R. If $z_1,\ldots,z_m$ is a finite subset of Ω_α then since Ω_α is a projective separable extension of $\Omega_{\alpha-1}$ by Theorem 2.1 there are elements $x_1,\ldots,x_n$ and $y_1,\ldots,y_n$ in Ω_α and $t \in \text{Hom}_{\Omega_{\alpha-1}}(\Omega_\alpha,\Omega_{\alpha-1})$ with

$$z_j = \Sigma_{i=1}^{n} x_i t(y_i z_j)$$

also

$$x_i x_j = \Sigma_{k=1}^{n} x_k t(y_k x_i x_j)$$

and

$$y_j = \Sigma_{i=1}^{n} x_i t(y_i y_j)$$

so let

$$A = \{t(y_k x_i x_j)\} \cup \{t(y_i y_j)\} \cup \{t(x_j)\} \cup \{t(y_i z_j)\}.$$

Observe that A is a finite subset of $\Omega_{\alpha-1}$ and since $\Omega_{\alpha-1}$ is locally separable over R there is a finite separable projective extension of S of R in $\Omega_{\alpha-1}$ containing A. Let T be the algebra over S generated by $x_1,\ldots,x_n$, let t' be the restriction of t to T. Observe that $t' \in \text{Hom}_S(T,S)$ and $y_1,\ldots,y_n$, $z_1,\ldots,z_m$ are elements of T. Applying Theorem 2.1 we see T is a finite projective separable extension of S and by Theorem 2.4 T is a finite projective separable extension of R. This proves Ω_α is a locally separable R-algebra. If α is a limit ordinal then the set of all ordinals less than α forms a linearly ordered set and employing Lemma 3.1 we let $\Omega_\alpha = \varinjlim_{\beta<\alpha} \Omega_\beta$. In this case one can check that Ω_α is a locally separable extension of R, that 0 and 1 are the only idempotents in Ω_α. If α and β are ordinals with $\alpha \neq \beta$ then $\Omega_\alpha \neq \Omega_\beta$.

Let S(R) be the collection of all finite projective separable extensions S of R with isomorphic extensions identified. Arguing as in Section 5 of Chapter 2 we observe the isomorphism classes of projective separable extensions of rank = n over R form a set so S(R) being the countable union of such sets is itself a set. Let λ be the ordinal whose cardinality is greater than

$$\sum_{S \in S(R)} \text{Rank}_R\ (S)\, |S|$$

where $|S|$ denotes the cardinality of S. By Corollary 1.6 there are at most Rank_R (S) isomorphisms of a finite projective separable extension S of R into Ω_λ. However since Ω_λ is locally separable we have

$$\Omega_\lambda = \cup\{S \mid S \subseteq \Omega_\lambda \text{ and } S \in S(R)\}.$$

Now for each S there are at most Rank_R (S) copies of S in the union, hence

$$\lambda \leq |\Omega_\lambda| \leq \Sigma_{S \in S(R)} \text{Rank}_R\ (S)\, |S| < \lambda$$

which is a contradiction. Therefore a separable closure of R must exist.

Next let S be a finite projective separable extension of R and let Ω be a separable closure of R. Then $\Omega \otimes S$ is a finite projective separable extension of Ω, so $\Omega \otimes S = \Omega \oplus \ldots \oplus \Omega$ with $n = \text{Rank}_R$ (S) summands in the decomposition. Let $\pi_1, \ldots, \pi_n$ be the Ω-algebra projections from $\Omega \otimes S$ to Ω given by this decomposition and let σ_i be the restriction of π_i to $1 \otimes S$. Then the σ_i are distinct R-algebra homomorphisms from S to Ω. If S has no idempotents other than 0 and 1 then by Corollary 2.6 each σ_i is a monomorphism. This proves the last statement of the theorem.

Let Ω be a separable closure of R and let Ω' be any locally separable extension of R. If S is a finite projective separable extension of R in Ω' then let G(S) be the set of all R-algebra homomorphisms from S to Ω. The set G(S) has Rank_R (S) elements and we make G(S) into a topological space by giving G(S) the discrete topology (point sets are open). In this topology G(S) is a compact Hausdorff space and if T is another finite projective separable extension of R in Ω' and $T \subseteq S$ then restriction gives a continuous function from G(S) to G(T). The collection of all finite projective separable extensions of R in Ω' forms a directed set under inclusion and we can form the inverse limit system of the topological spaces $\{G(S): S \in D\}$. By Lemma 3.2 $\overleftarrow{\lim}\, G(S) \neq \emptyset$. Using this fact we will construct an algebra homomorphism from Ω' to Ω. If $\overline{\sigma} \in \overleftarrow{\lim}\, G(S)$ and if $x \in \Omega'$ then there is a finite separable projective extension S of R in Ω' with $x \in S$ and if σ is the image of the projection of $\overline{\sigma}$ on G(S) then we define $h(x) = \sigma(x)$. Observe that the correspondence $\overline{\sigma} \to h$ is a one-to-one correspondence between $\overleftarrow{\lim}\, G(S)$ and the set of all R-algebra homomorphisms from Ω' to Ω.

If h is an R-algebra homomorphism from Ω' to Ω and $h(x) = 0$ then $x \in S$ for some finite projective separable extension S of R so since the restriction of h to S must be a monomorphism we must have $x = 0$. Thus every algebra homomorphism from Ω' to Ω is a monomorphism. If Ω' is also a separable closure of R then there are algebra monomorphisms $h: \Omega' \to \Omega$ and $g: \Omega \to \Omega'$. Now $\Omega'' = hg(\Omega)$ is a locally separable extension of R in Ω. If $x \in \Omega$ then there is a finite projective separable extension S of R in Ω containing x. There are Rank_R (S) monomorphisms of S into Ω and Rank_R (S) monomorphisms of S into Ω'' since hg is 1-1, so we must have $S \subseteq \Omega''$ and $x \in \Omega''$. Thus h is an isomorphism and Ω is unique.

We now obtain some results on the automorphisms of the separable closure.

Corollary 3.4: Let Ω be the separable closure of R, then any R-homomorphism from Ω to Ω is an automorphism of Ω. Moreover if S is a finite projective separable extension of R in Ω and σ is an R-homomorphism from S to Ω then there is an automorphism $\overline{\sigma}$ of Ω with the restriction of $\overline{\sigma}$ to S giving σ.

Proof: The first assertion of the corollary was proved while showing the uniqueness statement of Theorem 3.3. Next view Ω as an S-algebra in the usual way and let Ω' be the S-algebra Ω with the operation of S defined by $s \cdot x = \sigma(s)x$ for all $s \in S$, $x \in \Omega$, (notice that one must distinguish the multiplication of elements of S in Ω' with the operation of S on Ω'). One can check that both Ω and Ω' are separable closures of S so by Theorem 3.3 there is an S-algebra isomorphism f from Ω to Ω'. Therefore

$$f(sx) = s \cdot f(x) = \sigma(s)\ f(x)$$

for all $x \in \Omega$, $s \in S$. In particular letting $x = 1$ we have $f(s) = \sigma(s)$ for all $s \in S$.

Corollary 3.5: Let Ω be the separable closure of R. If $x \in \Omega - R$ there is an R-automorphism σ of Ω with $\sigma(x) \neq x$.

Proof: Let S be a finite projective separable extension of R in Ω containing x. By Theorem 2.9 S is contained in a normal separable extension N of R so that N has no idempotents other than 0 and 1. By Theorem 3.3, N may be viewed as a subalgebra of Ω containing S. There is an R-automorphism σ of N with $\sigma(x) \neq x$, and by Corollary 3.4 σ can be extended to an R-automorphism of Ω. This completes the proof.

With the results we have obtained thus far it is routine to generalize the infinite Galois theory of fields to the context of commutative rings whose only idempotents are 0 and 1. We will state the relevant definitions and quote the main theorem, leaving it to the reader to assemble the facts which have preceded to complete the proof.

If R is our commutative ring with no idempotents other than 0 and 1 and Ω is the separable closure of R we let G(R) be the full group of R-algebra automorphisms of Ω. We have seen in the proof of Theorem 3.3 that $G(R) = \overleftarrow{\lim}\, G(S)$ where S ranges over the finite projective separable extensions of R in Ω. Thus G(R) is a compact Hausdorff topological space and one can check that the operation of composition in G(R) is continuous so G(R) is a topological group. If S is an extension of R in Ω and $H = \{\sigma \in G(R) \mid \sigma(x) = x \text{ for all } x \in S\}$ then one can check that H is closed if and only if S is locally separable and that H is a normal subgroup of G(R) if and only if S is a normal extension of R.

Theorem 3.6: (Fundamental Theorem of Infinite Galois Theory)
Let R be a commutative ring whose only idempotents are 0 and 1 and let Ω be the separable closure of R. Let G(R) be the group of all R-algebra automorphisms of Ω. Then G(R) is a compact Hausdorff topological group and there is a one-to-one correspondence

$$H \to \Omega^H \qquad S \to \{\sigma \in G(R) \mid \sigma(x) = x \text{ for all } x \in S\}$$

between the closed subgroups H of G(R) and the locally separable extensions S of R in Ω. Moreover H is a closed normal subgroup of G(R) if and only if S is a normal extension of R, in this case $G(R)/H$ is the full group of automorphisms of S. Also S is a finite extension of R if and only if H is a subgroup of finite index in G(R) and in this case $\mathrm{Rank}_R\,(S) = [G:H]$.

One last remark may be of interest. We have associated with each commutative ring R with no idempotents other than 0 and 1 a unique compact Hausdorff topological group G(R) (unique by Theorem 3.3). One can show that this correspondence is a contravariant functor from the category of commutative rings with no idempotents other than 0 and 1 (and ring homomorphisms) into the category whose objects are compact Hausdorff topological groups and whose morphisms are described as follows: A morphism from the group G to the group H is an equivalence class of topological homomorphisms from G to H where two homomorphisms g, h are equivalent if there is an $y \in H$ so that $g(x) = y^{-1}h(x)y$ for all $x \in G$. What is involved is showing that for each pair R, K of commutative rings with no idempotents other than 0 and 1 there is a natural group homomorphism F which associates to any ring homomorphism f from R to K an equivalence class of continuous homomorphisms F(f) from G(K) to G(R). The homomorphism F is defined as follows: Let Ω be the separable closure of R and let Γ be the separable closure of K. Given f a ring homomorphism from R to K and using the observations that K is an R-algebra via f and $K \otimes \Omega$ is a locally separable K-algebra, one can show as in Theorem 3.3 that there is an extension of f to a homomorphism $\bar{f}$ from Ω to Γ. Given the extension $\bar{f}$ of f, if $\tau \in G(K)$ then $\tau \cdot \bar{f}$ is also an extension of f. Using Corollary 1.6 one can show that there is a unique element $\sigma \in G(R)$ such that $\tau \cdot \bar{f} = \bar{f} \cdot \sigma$. The equivalence class F(f) has a representative defined by $F(f)(\tau) = \sigma$. The different extensions of f to Ω yield different (but equivalent) homomorphisms from G(K) to G(R). It is now a straightforward matter to verify our association $R \to G(R)$ defines a functor from the category of commutative rings whose only idempotents are 0 and 1 to the category of compact Hausdorff topological groups.

§ 4. Separable Polynomials

Up to this point in Chapter 3 we have been dealing with the formal part of the Galois theory. In this section we present a generalization of the notions of "separable polynomial" and "separable element" to commutative rings. As in the previous section R will always be a commutative ring whose only idempotents are 0 and 1.

Definitions: Let R[x] be the ring of polynomials in one variable over R, and if $f(x) \in R[x]$ let $(f(x))$ denote the principal ideal in R[x] generated by f(x). If S is an R-algebra and $f(x) \in R[x]$ then the element $a \in S$ is called a root of f(x) in case $f(a) = 0$. A monic polynomial $f(x) \in R[x]$ is called <u>separable</u> in case $R[x]/(f(x))$ is a separable R-algebra. If S is an R-algebra and $b \in S$ then b is called a <u>separable element</u> in S in case b is the root of a separable polynomial $f(x) \in R[x]$. Notice that if f(x) is a separable polynomial then $R[x]/(f(x)) = S$ is a free separable extension of R and $\mathrm{Rank}_R\ (S)$ is the degree of the polynomial f(x). If S is an R-algebra and $b_1,\dots,b_m$ are elements of S we let $R[b_1,\dots,b_m]$ denote the R-subalgebra of S generated by $b_1,\dots,b_m$.

<u>Lemma 4.1:</u> Let S be a finite projective separable extension of R and assume 0 and 1 are the only idempotents in S. Let $b \in S$. Then b is a separable element in S if and only if R[b] is a separable extension of R in S.

<u>Proof</u>: If b is a separable element in S then there is a separable polynomial $f(x) \in R[x]$ with $f(b) = 0$. Thus R[b] is a homomorphic image of $R[x]/(f(x))$ so by Corollary 2.1.11 R[b] is separable over R.

To prove the converse we can assume by Theorem 2.9 that S is also a normal extension R with Galois group G. If R[b] is a separable subalgebra of S then by Theorem 1.1 there is a subgroup H of G with $S^H = R[b]$. Moreover S^H is a projective extension of R and $\mathrm{Rank}_R\ (S^H) = [G:H]$. Let $\sigma_1,\dots,\sigma_n$ be left coset representatives of H in G and let $f(x) = \prod_{i=1}^{n} (x - \sigma_i(b))$, then f(x) is a monic polynomial in R[x] of degree n and $f(b) = 0$. To see that f(x) is separable observe that $R[x]/(f(x))$ is a free R-module of rank n and the natural algebra homomorphism $h\colon R[x]/(f(x)) \to R[b]$ defined by $h(x) = b$ yields because R[b] is a projective R-module the split exact sequence

$$0 \to \text{kernel } (h) \to R[x]/(f(x)) \xrightarrow{h} R[b] \to 0.$$

Since the sequence is split and $\mathrm{Rank}_R\ (R[b]) = [G:H] = \mathrm{Rank}_R\ (R[x]/(f(x))$, we have $\mathrm{Rank}_R\ (\text{kernel } (h)) = 0$ so h is an isomorphism.

If S is a finite projective separable extension of R and if 0 and 1 are the only idempotents of S then we have shown in the proof of Lemma 4.1 that R[b] is a free R-module if b is a separable element in S.

<u>Theorem 4.2</u>: Let S be a finite normal separable extension of R with 0 and 1 the only idempotents of S and let f(x) be a separable polynomial in R[x]. If $b_1,\ldots,b_m$ are all the roots of f(x) in S then $R[b_1,\ldots,b_m]$ is a normal separable extension of R and $m \leq$ degree (f(x)).

<u>Proof</u>: Since $R[b_1,\ldots,b_m]$ is a homomorphic image of $R(b_1) \otimes \ldots \otimes R(b_m)$ and each $R(b_i)$ is separable over R by Lemma 4.1, we have that $R[b_1,\ldots,b_m]$ is separable over R. Let G be the Galois group of S over R. By Theorem 1.1 there is a subgroup H of G with $R[b_1,\ldots,b_m] = S^H$. Showing that S^H is a normal extension of R is equivalent to showing that $\sigma(R[b_1,\ldots,b_m]) \subseteq R[b_1,\ldots,b_m]$ for all $\sigma \in G$. But observe that if $\sigma \in G$ then the image of any root of f(x) under σ is another root of f(x). Thus $\sigma(R[b_1,\ldots,b_m] \subseteq R[b_1,\ldots,b_m]$ for all $\sigma \in G$ and $R[b_1,\ldots,b_m]$ is a normal separable extension of R. By Corollary 1.6 there are at most n distinct algebra homomorphisms from $R[x]/(f(x))$ to S where $n = \mathrm{Rank}_R\ (R[x]/(f(x)) =$ degree (f). The correspondences $x \to b_i$ define a set of m distinct algebra homomorphisms from $R[x]/(f(x))$ to S so $m \leq n$.

It is time to work toward a characterization of the separable polynomials in R[x]. Two ideals I_1 and I_2 of a commutative ring A are said to be <u>co-maximal</u> in case the ideal generated by the elements in $I_1 \cup I_2$ is A.

<u>Lemma 4.3</u>: Let S be a commutative R-algebra and assume 0 and 1 are the only idempotents in S. Let $f(x) \in R[x]$ be a separable polynomial and let a_1, a_2 be distinct roots of f(x) in S, then a_1-a_2 is invertible in S.

Proof: Observe that $S \otimes R[x]/(f(x)) = S[x]/(f(x))$ is a separable S-algebra. Let h be the natural homomorphism from $S[x]$ to $S[x]/(f(x))$, and let h_i be the homomorphism from $S[x]$ to S defined by $h_i(x) = a_i$. Since $(x-a_i)$ divides $f(x)$ in $S[x]$, h_i induces an S-algebra homomorphism $h_i'\colon S[x]/(f(x)) \to S$ so that $h_i = h_i' h$. By Lemma 1.5 there are orthogonal idempotents e_1, e_2 in $S[x]/(f(x))$ with $h_i'(e_j) = \delta_{i,j}$ and $[x + (f(x))]\, e_i = a_i e_i$ for $i,j = 1,2$. Applying h_i' if $p(x) \in S[x]$ and $[p(x) + (f(x))]e_i = 0$ we must have $p(a_i) = 0$. Therefore h maps the ideal $(x-a_i)$ of $S[x]$ onto the annihilator of e_i in $S[x]/(f(x))$ for $i = 1,2$. In particular this implies that $h(x-a_i) = S[x]/(f(x))\cdot(1-e_i)$ so by the orthogonality of e_1 and e_2 the ideals $h(x-a_1)$ and $h(x-a_2)$ are pairwise co-maximal in $S[x]/(f(x))$. Since $(x-a_i)$ contains $(f(x))$ $i = 1,2$ the ideals $(x-a_1)$ and $(x-a_2)$ are pairwise co-maximal in $S[x]$. Thus there are polynomials $q(x)$ and $r(x)$ in $S[x]$ with

$$q(x)\ (x-a_1) + r(x)\ (x-a_2) = 1.$$

Letting $x = a_1$ we have $r(a_1)(a_1-a_2) = 1$ so a_1-a_2 is invertible in S.

We can now characterize the separable polynomials over R.

Theorem 4.4: Let $f(x)$ be a monic polynomial in $R[x]$, then the following statements are equivalent.

1) $f(x)$ is a separable polynomial.
2) There is a finite projective separable extension S of R with no idempotents other than 0 and 1 in S so that S contains elements $a_1,\ldots,a_n$ such that a_i-a_j is invertible in S when $i \neq j$ and $f(x) = (x-a_1)\ \ldots\ (x-a_n)$.
3) Let t denote the trace map of the free R-module $R[x]/(f(x))$ and let $y = x + (f(x))$. If n = degree $(f(x))$ and if we let $[t(y^i y^j)]$ be the $n \times n$ matrix whose $i + 1$, $j + 1$ entry is $t(y^i y^j)$ then the determinant of $[t(y^i y^j)]$ is invertible in R.

Proof: 1) implies 2). By Lemma 2.8 there is a finite projective separable extension S of R so that 0 and 1 are the only idempotents in S and so that $S \otimes R[x]/(f(x)) \cong S \oplus \ldots \oplus S$ where there are n = degree f(x) copies of S in the decomposition. Identify $S \otimes R[x]/(f(x))$ with $S[x]/(f(x))$, then we have $S[x]/(f(x)) = Se_1 \oplus \ldots \oplus Se_n$ for suitable idempotents $e_1, \ldots, e_n$ in $S[x]/(f(x))$. If $[x + (f(x))]\, e_i = a_i e_i$ with $a_i \in S$ then $f(a_i) = 0$. The correspondences $[x + (f(x))] \to a_i$ provide distinct S-algebra homomorphisms from $S[x]/(f(x))$ to S so if $i \neq j$ then $a_i \neq a_j$. By Lemma 4.3, $a_i - a_j$ is invertible in S if $i \neq j$ and f(x) has n-roots in S. We can find a polynomial $p_1(x) \in S[x]$ with $f(x) = (x-a_1)\, p_1(x)$. Since $a_1 - a_2$ is invertible and $f(a_2) = 0$ we must have $p_1(a_2) = 0$ so $p_1(x) = (x-a_2)\, p_2(x)$. Continue in this way until $f(x) = (x-a_1)(x-a_2) \ldots (x-a_n)$.

2) implies 1). Let S have the properties stated in 2), then the ideals $(x-a_i)$ are pairwise co-maximal in S[x] for $i = 1, \ldots, n$ since the ideal containing $x-a_i$ and $x-a_j$ contains the unit $a_i - a_j$. Also the intersection of the ideals $(x-a_i)$ is (f(x)). Therefore by the Chinese Remainder Theorem

$$S[x]/(f(x)) = S \oplus \ldots \oplus S. \qquad \text{(n-copies of S)}$$

Hence $S \otimes R[x]/(f(x))$ is separable over S so by Corollary 2.3 and Proposition 2.1.10 $R[x]/(f(x))$ is separable over R.

1) equivalent to 3). Observe that $R[x]/(f(x))$ is a free R-module with basis $\{1, y, \ldots y^{n-1}\}$ where n = degree f(x). Let π_i be the R-projection of $R[x]/(f(x))$ onto the coefficients of y^i. One can check that the trace map t is given by the equation

$$t(z) = \Sigma_{i=0}^{n-1}\, \pi_i(zy^i) \qquad z \in R[x]/(f(x)).$$

From now on in the proof let T denote $R[x]/(f(x))$. T will be separable over R if and only if $\mathrm{Hom}_R\,(T,R)$ is generated as a right T-module by the trace map t. Thus T is separable over R if and only if there are elements $z_i \in T$ with $t(z_i -) = \pi_i$. Suppose T is separable and the z_i

exist. Let $z_i = \sum_{j=0}^{n-1} a_{ij}y^j$ where $a_{ij} \in R$. Since $t(z_i -) = \pi_i$

$$\pi_i(y^j) = \delta_{i,j} = t(z_i y^j) = \sum_{j=0}^{n-1} a_{ij}\ t(y^i y^j).$$

In matrix notation $(a_{ij})(t(y^i y^j)) = I_n$. Thus the determinant of $t(y^i y^j)$ is invertible in R. These steps are reversible so the proof of the theorem is complete.

Using condition 3 of Theorem 4.4 one can verify that the polynomial $x^n - b \in R[x]$ is separable if and only if $n \cdot 1$ and b are invertible in R. Thus by Theorem 4.2, if $n \cdot 1$ is invertible in R there are at most n distinct n^{th} roots of 1 in R. This need not be the case in general. For example, the group ring of the abelian group of order two over the integers is a commutative ring in which there are more than 2 distinct square roots of 1. This commutative ring also has no idempotents other than 0 and 1.

If S is a field which is a finite separable extension of the field R then the "Primitive Element Theorem" asserts that there is a separable element $a \in S$ so that $S = R[a]$. It is at this point that the general Galois theory in commutative rings fails to keep pace with the classical results. For most classes of commutative rings there appears to be no analogue of the Primitive Element Theorem available. It is for this reason that the separable polynomials cannot play the central role in the Galois theory of commutative rings as they do in the field theory. On p. 170 of E. Weiss' book "Algebraic Number Theory" one can find an example due to Dedekind of a local principal ideal domain R and a finite projective separable extension S over R so that S contains no idempotents other than 0 and 1 and such that no element in S-R is separable over R.

Exercises

1. Show t_M is independent of the choice of dual basis for M (See III-13).

2. Let T be an R-separable subalgebra of the finite projective separable extension S of R. Show T is finite and projective over R.

3. Let $R = Z/(p^n)$ and let S be a finite projective separable extension of R. Show there is an $a \in S$ with $S = R[a]$.

4. In analogy with the theory for finite fields, classify all finite separable projective extensions of $Z/(p^n)$.

5. (Hongan and Nagahara [55]). Let S be a finite projective separable extension of R and assume the only idempotents in S are 0 and 1. Show S is a field if and only if R[a] is separable over R for any $a \in S$.

6. Let N be a finite normal separable extension of the field R. Assume 0 and 1 are the only idempotents in N. Show N is a field.

7. Let K be the ring of continuous real valued functions f(x) on the closed interval $0 \leq x \leq 1$ so that $f(0) = f(1)$. Let G be a finite group of automorphisms of K and $R = K^G$. Prove that K is a normal separable extension of R. Show that for any positive integer n there is a group of automorphisms of K whose order is equal to n.

8. Let R be a commutative ring with no idempotents other than 0 and 1. Let f(x) be a monic polynomial in R[x]. The following are equivalent.

   1. f(x) is a separable polynomial over R.
   2. For each maximal ideal $\mathfrak{M}$ of R, f(x) is separable when viewed as a polynomial over $R_{\mathfrak{M}}$.
   3. For each maximal ideal $\mathfrak{M}$ of R, the polynomial obtained from f(x) by reducing the coefficients modulo $\mathfrak{M}$ does not have repeated roots in an algebraic closure of $R/\mathfrak{M}$.

CHAPTER IV

In this chapter we derive a six term exact sequence which reduces to the two classical theorems in Galois cohomology of fields, Hilbert's Theorem 90 and the "Crossed Product" Theorem. The method of proof is to exhibit the natural homomorphisms connecting the terms in the sequence and prove the exactness directly. Extensive use is made of the results and techniques of the previous chapters.

§ 1. Definitions and Statement of the Theorem

Let G be a finite group and let M be an abelian group on which G acts as a group of operators. It will usually be convenient to view the operation in both G and M as multiplication. Let

$$Z^1(G,M) = \{f:G \to M \mid f(\sigma\tau) = f(\sigma)\sigma\cdot f(\tau) \text{ for all } \sigma,\tau \in G\}$$

$$B^1(G,M) = \{f \in Z^1(G,M) \mid \text{there exists } m\in M \text{ with } f(\sigma)=\sigma(m)/m \text{ for all } \sigma\in G\}$$

$$Z^2(G,M) = \{f:G\times G \to M \mid f(\sigma\tau,\rho)\ f(\sigma,\tau) = f(\sigma,\tau\rho)\sigma f(\tau,\rho) \text{ for all } \sigma,\tau,\rho \in G\}$$

$$B^2(G,M) = \{f\in Z^2(G,M) \mid \text{there exists } g:G \to M \text{ with } f(\sigma,\tau)=g(\sigma\tau)/[g(\sigma)\sigma\cdot g(\tau)] \text{ for all } \sigma,\tau \in G\}.$$

Observe that $Z^i(G,M)$ is an abelian group under pointwise multiplication and that $B^i(G,M)$ is a subgroup $(i=1,2)$. Let $H^i(G,M) = Z^i(G,M)/B^i(G,M)$, $H^i(G,M)$ is called the ith <u>cohomology group</u> of G with coefficients in M $(i = 1,2)$. If $f \in Z^i(G,M)$ the class f represents in $H^i(G,M)$ will be denoted $[f]$. Recall that if S is a commutative ring we have let P(S) be the group of rank-one projective S-modules and U(S) be the group of units in S. If S is an extension of R we have let B(S/R) be the subgroup of B(R) consisting of the elements in B(R) which are split by S. If G is a finite group of R-automorphisms of S then both U(S) and P(S) are G-modules. The action of G on U(S) is the obvious one and if $[E] \in P(S)$ and $\sigma \in G$ then $\sigma\cdot[E] = [E_\sigma]$ where E_σ is isomorphic to E as an R-module and if $x \in E_\sigma$, $s \in S$ then $s\cdot x = \sigma(s)x$. We let $P(S)^G$ consist of the subgroup of elements $[E] \in P(S)$ with $\sigma\cdot[E] = [E]$ for all $\sigma \in G$. The principal theorem of this chapter can now be stated.

Theorem 1.1: Let R be any commutative ring and let S be a Galois extension of R with Galois group G. Then there are natural homomorphisms α_i which give the exact sequence

$$1 \to H^1(G,U(S)) \xrightarrow{\alpha_1} P(R) \xrightarrow{\alpha_2} P(S)^G \xrightarrow{\alpha_3} H^2(G,U(S)) \xrightarrow{\alpha_4} B(S/R) \xrightarrow{\alpha_5} H^1(G,P(S)).$$

There are two important corollaries of the theorem.

Corollary 1.2: (Hilbert's Theorem 90) If $P(R) = \{1\}$ then $H^1(G,U(S)) = \{1\}$.

Corollary 1.3: (Crossed Product Theorem) If $P(S) = \{1\}$ then $B(S/R) \simeq H^2(G,U(S))$.

§ 2. The Proof of Theorem 1.1

The remainder of this chapter is devoted to a proof of Theorem 1.1. We will define in turn each of the homomorphisms α_i and prove exactness at each term as we go.

2.1 Preliminaries to the Definition of α_1: Since S is a Galois extension of R with group G we know by Proposition 3.1.2 that $\Delta(S:G) \simeq \mathrm{Hom}_R(S,S)$ under map j and that this puts us in the context of the Morita Theorems. If M is a left $\Delta(S:G)$-module then M is a left G-module with action defined by $\sigma\cdot M = u_\sigma m$ for all $m \in M$, $\sigma \in G$, $u_\sigma \in \Delta$. Let $M^G = \{m \in M \mid \sigma\cdot m = m \text{ for all } \sigma \in G\}$. There is a map $\psi\colon S \otimes M^G \to M$ defined by $\psi(s \otimes m) = sm$. If we view $S \otimes M^G$ as a left $\Delta(S:G)$-module by letting $\Delta(S:G)$ act on the first variable then ψ is a left $\Delta(S:G)$-module homomorphism. By Corollary 3.2.2, $\mathrm{Hom}_R(S,R) = t(S-)$ where $t(x) = \sum_{\sigma\in G} \sigma(x)$ for all $x \in S$. Let $\Delta = \Delta(S:G)$ and let $\phi\colon \mathrm{Hom}_R(S,R) \otimes_\Delta M \to M^G$ be defined by $\phi(t(s-) \otimes m) = t(sm)$ for all $s \in S$, $m \in M$. By Corollary 3.1.3 there is an element $c \in S$ with

$t(c) = 1$ so if $m \in M^G$ then $m = t(cm) = \phi(t(c-) \otimes m)$. Therefore ϕ is an R-epimorphism from $\mathrm{Hom}_R(S,R) \otimes_\Delta M$ onto M^G. We have the commuting diagram

$$\begin{array}{ccc} S \otimes M^G & \xrightarrow{\quad \psi \quad} & M \\ {\scriptstyle 1 \otimes \phi} \nwarrow & & \nearrow {\scriptstyle \gamma} \\ & S \otimes (\mathrm{Hom}_R(S,R) \otimes_\Delta M) & \end{array}$$

The $\Delta(S:G)$ isomorphism γ is given in Morita's Theorem (Prop. 1.3.3), $\gamma(s_1 \otimes (t(s_2-) \otimes m) = s_1 \sum_{\sigma\in G} \sigma\cdot(s_2 m)$. Since $1 \otimes \phi$ is an epimorphism both ψ and $1 \otimes \phi$ are isomorphisms. Since ϕ is the restriction of $1 \otimes \phi$, ϕ is an isomorphism.

2.2 <u>The definition of α_1 and the proof that α_1 is a well defined group homomorphism</u>:

Given $f \in Z^1(G,U(S))$ we let $\theta_f:\Delta(S:G) \to \Delta(S:G)$ be defined by $\theta_f(su_\sigma) = s\, f(\sigma)u_\sigma$. Observe that θ_f is an algebra homomorphism so by Theorem 2.6.1 θ_f is an automorphism of $\Delta(S:G)$. One can also check that $\theta_{fg} = \theta_f\theta_g$. If $f \in B^1(G,U(S))$ then $f(\sigma) = \sigma(a)/a$ for some $a \in U(S)$ and all $\sigma \in G$, thus $\theta_f(su_\sigma) = a^{-1}(su_\sigma)a$. Define a $\Delta(S:G)$-module S_f by letting $\Delta(S:G)$ act on S by the rule $(su_\sigma)\cdot x = \theta_f(su_\sigma)x$ for all x, $s \in S$ and $\sigma \in G$. As we just saw there is a $\Delta(S:G)$-isomorphism

$$S_f \simeq S \otimes S_f^G.$$

The rank of both S_f and S are $[G:1]$ over R so $|S_f^G| \in P(R)$. If $f \in B^1(G,U(S))$ then $\theta_f(su_\sigma) = a^{-1}(su_\sigma)\, a$ for some $a \in U(S)$ so there is a R-isomorphism from R to S_f^G given by $r \to a^{-1}r$. In this case $|S_f^G| = |R|$. For any f, $g \in Z^1(G,U(S))$ there is a chain of natural S-isomorphisms whose composition is a $\Delta(S:G)$-isomorphism.

$$S \otimes (S_f^G \otimes S_g^G) \simeq (S \otimes S_f^G) \otimes_S (S \otimes S_g^G) \simeq S_f \otimes_S S_g \simeq S_{fg} \simeq S \otimes S_{fg}^G.$$

Define $\alpha_1([f]) = |S_f^G|$. By Proposition 1.3.3 we have shown that α_1 is a well defined homomorphism from $H^1(G,U(S))$ to $P(R)$.

2.3 Exactness at $H^1(G,U(S))$: Assume $\alpha_1\ [f] = |S_f^G| = |R|$. Then as we have seen, $S_f \cong S \otimes S_f^G \cong S$ over $\Delta(S:G)$. Let ω: $S_f \to S$ be the $\Delta(S:G)$-isomorphism. As R-modules $S_f \cong S$ so $\omega \in \mathrm{Hom}_R\ (S,R) \cong \Delta(S:G)$. Thus there is an element $w = \Sigma_{\tau \in G}\beta_\tau u_\tau \in \Delta(S:G)$ so that $\omega(x) = w\cdot x$ for all $x \in S$. Now $v\omega(x) = \omega(\theta_f(v)x)$ for all $v \in \Delta(S:G)$ and $x \in S$ so $v\cdot w = w\theta_f(v)$ for all $v \in \Delta(S:G)$. Since ω is an isomorphism, w is invertible and

$$\theta_f(v) = w^{-1}vw \text{ for all } v \in \Delta(S:G).$$

After identifying S with Su_1 in $\Delta(S:G)$ we will show $w \in U(S)$ and then reverse the procedure used in defining α_1 to show $f \in B^1(G,U(S))$. Let $v = su_1$, then $\theta_f(su_1) = su_1 = w^{-1}su_1w$. Thus

$$\Sigma_{\tau \in G}\beta_\tau \tau(s)u_\tau = \Sigma_{\tau \in G}\beta_\tau s u_\tau$$

and so $\beta_\tau\tau(s) = \beta_\tau s$ for all $\tau \in G$ and all $s \in S$. Thus $\beta_\tau u_\tau - \beta_\tau u_1 = 0$ in $\Delta(S:G)$ so $\beta_\tau = 0$ unless $\tau = 1$. We can therefore identify $w = \beta_1 u_1$ with the element β_1 in $U(S)$. Finally,

$$\theta_f(u_\tau) = f(\tau)u_\tau = \beta_1^{-1}\ u_\tau\beta_1 = \beta_1^{-1}\tau(\beta_1)u_\tau$$

therefore $f(\tau) = \tau(\beta_1)/\beta_1$ for all $\tau \in G$ so $f \in B^1(G,U(S))$.

2.4 Definition of the Homomorphism α_2: Recalling how $P(S)$ was made into a G-module we observe that if $|E| \in P(R)$ then $\sigma\cdot(S \otimes E) \simeq S \otimes E$ by the map $\sigma \otimes 1$ so $|S \otimes E| \in P(S)^G$. Define $\alpha_2(E) = |S \otimes E|$ for all $|E| \in P(R)$. It is now clear that α_2 is a homomorphism from $P(R)$ to $P(S)^G$.

2.5 Exactness at $P(R)$: Let $f \in Z^1(G,U(S))$, then $\alpha_2\alpha_1\ [f] = \alpha_2|S_f^G| = |S \otimes S_f^G|$ and $S \otimes S_f^G \simeq S_f$. The last isomorphism is as $\Delta(S:G)$-modules. Thus $S_f \simeq S$ as S-modules and $\alpha_2\alpha_1 = 1$ so image $\alpha_1 \subseteq$ kernel α_2.

Let $|E| \in P(R)$ and assume $|E| \in$ kernel α_2. Then there is an S-module isomorphism ϕ: $S \otimes E \to S$. Turn $S \otimes E$ into a $\Delta(S:G)$-module by letting $\Delta(S:G)$ act on the first variable. Identifying $\Delta(S:G)$ with $\mathrm{Hom}_R\ (S,S)$ define the map $\theta:\Delta(S:G) \to \Delta(S:G)$ by $\theta(w)\cdot x = \phi(w\cdot\phi^{-1}(x))$

for all $w \in \Delta(S:G)$ and $x \in S$. Observe that θ is an R-algebra and S-module isomorphism, so $\theta(s) = s$ for all $s \in S$ and θ is determined by its action on the basis elements u_σ of $\Delta(S:G)$. Let $f(\sigma) = \theta(u_\sigma)\cdot 1$ (where $\theta(u_\sigma)$ is viewed as an element in $\mathrm{Hom}_R\,(S,S)$). Since $\theta(u_\sigma)$ is an isomorphism, and since $\theta(u_\sigma)\cdot s = \sigma(s)\ f(\sigma)$ we have that $f(\sigma) \in U(S)$ and $\theta(u_\sigma) = f(\sigma)u_\sigma$ for all $\sigma \in G$. Thus θ is determined by the function f and since $\theta(u_\sigma)\theta(u_\tau) = \theta(u_{\sigma\tau})$ we have $f(\sigma\tau) = f(\sigma)\sigma\cdot f(\tau)$ so $f \in Z^1(G,U(S))$. The proof will be complete when we exhibit an R-module isomorphism from E to S_f^G. Now $S \otimes E \simeq S$ under the isomorphism ϕ and $\phi u_\sigma \phi^{-1} = f(\sigma)\ u_\sigma$ so if $z \in S_f$ and $su_\sigma \in \Delta(S:G)$ then $su_\sigma\cdot z = sf(\sigma)\sigma(z)$. Thus

$$\begin{aligned}\phi^{-1}(su_\sigma\cdot z) &= \phi^{-1}(sf(\sigma)\ u_\sigma\cdot z)\\ &= \phi^{-1}(\phi su_\sigma\phi^{-1}\cdot z)\\ &= su_\sigma\phi^{-1}(z).\end{aligned}$$

Therefore ϕ^{-1} is an $\Delta(S:G)$-isomorphism from S_f to $S \otimes E$ and $|E| = |S_f^G|$. Thus $|E| \in$ image α_1 which proves exactness at P(R).

2.6 <u>The Definition of α_3</u>: If $|V| \in P(S)^G$ then $|\sigma\cdot V| = |V|$ for each $\sigma \in G$. Therefore there is an R-isomorphism ψ_σ: $V \to V$ such that $\psi_\sigma(sx) = \sigma(s)\psi_\sigma(x)$ for all $x \in S$, $x \in V$. The map $\psi_{\sigma\tau}\psi_\tau^{-1}\psi_\sigma^{-1}$ is an R-isomorphism from V to V and $\psi_{\sigma\tau}\psi_\tau^{-1}\psi_\sigma^{-1}(sx) = s\psi_{\sigma\tau}\psi_\tau^{-1}\psi_\sigma^{-1}(x)$ for all $s \in S$, $x \in V$. Therefore $\psi_{\sigma\tau}\psi_\tau^{-1}\psi_\sigma^{-1} \in \mathrm{Hom}_S\,(V,V) = S$ and $\psi_{\sigma\tau}\ \psi_\tau^{-1}\psi_\sigma^{-1}$ is multiplication by an element $f(\sigma,\tau)$ in U(S). With each element $V \in P(S)^G$ we have associated a function $f: G\times G \to U(S)$. From the equation $\psi_{\sigma(\tau\rho)} = \psi_{(\sigma\tau)\rho}$ one can check that $f \in Z^2(G,U(S))$. We define α_3 by the correspondence $|V| \to [f]$.

2.7 <u>α_3 is well defined</u>: Let $|V| \in P(S)^G$ and define the element $f \in Z^2(G,U(S))$ as above. Let $\{\lambda_\sigma | \sigma \in G\}$ be other choices for R-isomorphisms from V to V with $\lambda_\sigma(sx) = \sigma(s)\lambda_\sigma(x)$ for all $s \in S$, $x \in V$. Let

$g:G \times G \to U(S)$ be defined by $g(\sigma,\tau) = \lambda_{\sigma\tau}\lambda_\tau^{-1}\lambda_\sigma^{-1}$ for all $\sigma, \tau \in G$. As before $g \in Z^2(G,U(S))$. Since $\lambda_\sigma\psi_\sigma^{-1}$ is an S-isomorphism from V to V we have $\lambda_\sigma\psi_\sigma^{-1}$ is multiplication by an element $h(\sigma) \in U(S)$.

$$\begin{aligned} f(\sigma,\tau)^{-1}g(\sigma,\tau) &= (\psi_\sigma\psi_\tau\psi_{\sigma\tau}^{-1})(\lambda_{\sigma\tau}\lambda_\tau^{-1}\lambda_\sigma^{-1}) \\ &= (\psi_\sigma\lambda_\sigma^{-1})\lambda_\sigma(\psi_\tau\lambda_\tau^{-1})\lambda_\tau\psi_{\sigma\tau}^{-1}\lambda_{\sigma\tau}\lambda_\tau^{-1}\lambda_\sigma^{-1} \\ &= (\sigma\cdot h(\tau)\ h(\sigma))^{-1}(\lambda_\sigma\lambda_\tau\psi_{\sigma\tau}^{-1})(\lambda_{\sigma\tau}\lambda_\tau^{-1}\lambda_\sigma^{-1}) \\ &= (\sigma\cdot h(\tau)\ h(\sigma))^{-1}\ h(\sigma\tau). \end{aligned}$$

Thus $[f] = [g]$ in $H^2(G,U(S))$ and α_3 is well defined.

2.8 <u>α_3 is a homomorphism</u>: Let $|V|, |W| \in P(S)^G$, and assume $\alpha_3|V| = [f]$ and $\alpha_3|W| = [g]$. For each $\sigma \in G$ let $\psi_\sigma:V \to V$ and $\theta_\sigma:W \to W$ be the R-isomorphisms which define f and g respectively. Then $\psi_\sigma \otimes \theta_\sigma:V \otimes_S W \to V \otimes_S W$ is an R-isomorphism with the required property to define an element $h \in Z^2(G,U(S))$. For all $\sigma, \tau \in G$

$$\begin{aligned} h(\sigma,\tau) &= (\psi_{\sigma\tau} \otimes \theta_{\sigma\tau})(\psi_\tau \otimes \theta_\tau)^{-1}(\psi_\sigma \otimes \theta_\sigma)^{-1} \\ &= \psi_{\sigma\tau}\ \psi_\tau^{-1}\psi_\sigma^{-1} \otimes \theta_{\sigma\tau}\ \theta_\tau^{-1}\theta_\sigma^{-1} \\ &= f(\sigma,\tau)\ g(\sigma,\tau). \end{aligned}$$

Thus $\alpha_3|V \otimes_S W| = [f]\ [g] = \alpha_3|V|\alpha_3|W|$ and α_3 is a homomorphism.

2.9 <u>Exactness at $P(S)^G$</u>: If $|V| \in$ image α_2 there is an $|E| \in P(R)$ with $|V| = |S \otimes E|$. In this case $V \simeq \sigma\cdot V$ by the map $\psi_\sigma(s \otimes x) = \sigma(s) \otimes x$ for all $s \in S$, $x \in E$ and for each $\sigma \in G$. Observe that $\psi_\sigma\psi_\tau = \psi_{\sigma\tau}$ and $\alpha_3|V| = [f]$ where $f(\sigma,\tau) = \psi_{\sigma\tau}\ \psi_\tau^{-1}\psi_\sigma^{-1} = 1$ for all $\sigma, \tau \in G$. Therefore, image $\alpha_2 \subseteq$ kernel α_3.

If $|V| \in$ kernel α_3 then for each $\sigma \in G$ there is an isomorphism $\psi_\sigma:V \to V$ and an element $h(\sigma) \in U(S)$ so that

$$\psi_\sigma(sx) = \sigma(s)\psi_\sigma(x) \quad \text{for all } s \in S,\ x \in V \quad \text{and}$$

$$\psi_{\sigma\tau}\psi_{\tau}^{-1}\psi_{\sigma}^{-1} = h(\sigma\tau)\ h(\sigma)^{-1}\sigma\cdot h(\tau)^{-1} \text{ for all } \sigma,\ \tau \in G.$$

Observe that V can be viewed as a left $\Delta(S:G)$-module by defining $u_\sigma \cdot x = h(\sigma)^{-1}\psi_\sigma(x)$ for all $\sigma \in G$ and $x \in V$. In 2.1 we showed $V \simeq S \otimes V^G$. By a rank argument $|V^G| \in P(R)$ so $|V| = \alpha_2 |V^G|$ and kernel α_3 = image α_2.

2.10 Preliminaries to the definition of α_4: If S is a field the "crossed product" theorem asserts the existence of a natural isomorphism from $H^2(G,U(S))$ to $B(S/R)$. One of the purposes of the six-term sequence we are deriving is to measure the distance by which this result may fail in our setting. The reader may find it instructive to consult the monograph of Artin, Nesbitt and Thrall or Volume II of "Modern Algebra" by Van der Waerden for a proof of the "crossed product" theorem when S is a field.

2.11 The definition of α_4: If $f \in Z^2(G,U(S))$ we can construct the R-algebra $\Delta(S:G:f)$ which is a free S-module with S-basis $\{u_\sigma | \sigma \in G\}$ and multiplication defined by $(s_\sigma u_\sigma)(s_\tau u_\tau) = s_\sigma\ \sigma(s_\tau)\ f(\sigma,\tau)\ u_{\sigma\tau}$ for all $\sigma, \tau \in G$; $s_\sigma, s_\tau \in S$. Notice that $\Delta(S:G) = \Delta(S:G:1)$. If $g \in Z^2(G,U(S))$ with $[g] = [f]$ then there is a map $h:G \to U(S)$ with $g(\sigma,\tau) = f(\sigma,\tau)h(\sigma\tau)h(\sigma)\sigma h(\tau)^{-1}$. In this case there is an R-algebra isomorphism from $\Delta(S:G:f)$ to $\Delta(S:G:g)$ defined by $s_\sigma u_\sigma \to s_\sigma h(\sigma) u_\sigma$. We let $\alpha_4[f] = \Delta(S:G:f)$ for all $[f] \in H^2(G,U(S))$.

2.12 α_4 is well defined: To verify that α_4 is well defined we must show $\Delta(S:G:f) \in B(S/R)$.

Observe, using ii of Proposition 3.1.2, that $S \otimes S$ can be viewed as a Galois extension of R with Galois group $G \times G$ where if $(\sigma,\tau) \in G \times G$ then $(\sigma,\tau)(x \otimes y) = \sigma(x) \otimes \tau(y)$ for all $x, y \in S$.

By Proposition 3.1.2 iv there is a set $\{e_\sigma | \sigma \in G\}$ of orthogonal idempotents in $S \otimes S$ defined by $e_\sigma = \ell^{-1}(v_{\sigma^{-1}})$. Using the isomorphism

ℓ one can check that for all $x \in S$; $\sigma, \tau, \rho \in G$

$$\text{i)} \quad (1 \otimes \sigma(x))e_\sigma = (x \otimes 1)e_\sigma$$

$$\text{ii)} \quad (\sigma, \tau)e_\rho = e_{\tau\rho\sigma^{-1}} .$$

If $f,g \in Z^2(G,U(S))$ then for any $\sigma, \tau, \rho, \eta \in G$ we define

$(f,1) \in Z^2(G\times 1, U(S \otimes S))$ by $(f,1)[(\sigma,1),(\tau,1)] = f(\sigma,\tau) \otimes 1$

$(f,g) \in Z^2(G\times G, U(S \otimes S))$ by $(f,g)[(\sigma,\tau),(\eta,\rho)] = f(\sigma,\eta) \otimes g(\tau,\rho)$.

Let $h: G \times 1 \to U(S \otimes S)$ be defined by

$$h(\sigma,1) = \Sigma_{\tau \in G} (1 \otimes f(\tau,\sigma))e_\tau .$$

Identifying $G \times 1$ with G we have that for any $\sigma, \tau \in G$

$$\begin{aligned}
h(\sigma\tau)h(\sigma)^{-1}\sigma h(\tau)^{-1} &= [\Sigma_{\rho \in G} 1\otimes f(\rho,\sigma\tau)e_\rho][\Sigma_{\beta \in G} 1\otimes f(\beta,\sigma)^{-1}e_\beta]\sigma[\Sigma_{\gamma \in G} 1\otimes f(\gamma,\tau)^{-1}e_\gamma] \\
&= [\Sigma_{\rho \in G} 1\otimes f(\rho,\sigma\tau)f(\rho,\sigma)^{-1}e_\rho][\Sigma_{\gamma \in G} 1\otimes f(\gamma,\tau)^{-1}e_{\gamma\sigma^{-1}}] \\
&= \Sigma_{\rho \in G} 1\otimes f(\rho,\sigma\tau)f(\rho,\sigma)^{-1}f(\rho\sigma,\tau)^{-1}e_\rho \\
&= \Sigma_{\rho \in G} 1\otimes f(\rho,\sigma\tau)f(\rho,\sigma\tau)^{-1}\rho f(\sigma,\tau)^{-1}e_\rho \\
&= \Sigma_{\rho \in G} (1\otimes\rho)\ f(\sigma,\tau)^{-1}e_\rho \\
&= \Sigma_{\rho \in G} (f(\sigma,\tau)^{-1} \otimes 1)e_\rho \\
&= f(\sigma,\tau)^{-1} \otimes 1 \\
&= (f,1)^{-1}[(\sigma,1),\ (\tau,1)].
\end{aligned}$$

Therefore $(f,1)^{-1} \in B^2(G,U(S \otimes S))$ so $(f,1) \in B^2(G,U(S \otimes S))$. Observe that $S \otimes S$ is a Galois extension of $1 \otimes S$ with group $G \times 1$ and that $S \otimes \Delta(S:G:f) \simeq \Delta(S \otimes S:G \times 1:\ (f,1))$ under the correspondence $s \otimes a_\sigma u_\sigma \to (a_\sigma \otimes s)u_{(\sigma,1)}$; $s, a_\sigma \in S$, $\sigma \in G$. Since $(f,1) \in B^2(G,U(S\otimes S))$ there is a function $h: G \to U(S \otimes S)$ so that for all $\sigma, \tau \in G$

$$f(\sigma,\tau) \otimes 1 = h(\sigma\tau)^{-1}\ h(\sigma)\sigma\cdot h(\tau).$$

In this case $\Delta(S \otimes S:G \times 1:\ (f,1)) \simeq \Delta(S \otimes S:\ G \times 1)$ under the correspondence $u_{(\sigma,1)} \to h(\sigma)\ u_{(\sigma,1)}$. We know by Proposition 3.1.2 iii that $\Delta(S \otimes S:\ G \times 1) = \mathrm{Hom}_S\ (S \otimes S,\ S \otimes S)$ is a central separable S-algebra in the zero-class of $B(S)$. Now $R\cdot 1$ is an R-direct summand of S by Corollary 3.1.3, we have shown $S \otimes \Delta(S:G:f) = \Delta(S \otimes S:G \times 1)$ is a central separable S-algebra so by Corollary 2.1.10 $\Delta(S:G:f)$ is a central separable R-algebra. This proves α_4 is well defined.

2.13 $\underline{\alpha_4}$ is a homomorphism: Saving the notation of the preceding section we must show that for any $[f]$, $[g] \in H^2(G,U(S))$,

$$\Delta(S:G:f) \otimes \Delta(S:G:g) = \Delta(S:G:fg) \text{ in } B(R).$$

Let f, $g \in Z^2(G,U(S))$ and let $(f,g) \in Z^2(G \times G,U(S \otimes S))$ be the element defined in 2.12. The principal step in showing α_4 is a homomorphism is showing

$$[(f,g)] = [(fg,1)] \text{ in } H^2(G \times G,\ U(S \otimes S)).$$

Observe that $(f,g)(g,g^{-1}) = (fg,1)$ so if $(g,g^{-1}) \in B^2(G \times G,\ U(S \otimes S))$ then $[(f,g)] = [(fg,1)]$. Also observe that there is a map ψ from $\Delta(S:G:g)^{\circ}$ to $\Delta(S:G:g^{-1})$ defined by $\psi(s_\sigma u_\sigma) = \sigma^{-1}(s_\sigma)(u_\sigma)^{-1}$. It is clear that ψ is an R-module isomorphism. If $u_\sigma \in \Delta(S:G:g^{-1})$, $a \in S$, $b \in U(S)$ then the three formulae i) $(u_\sigma)^{-1} = \sigma^{-1}(g(\sigma,\sigma^{-1})u_{\sigma^{-1}}$

ii) $(u_\sigma)^{-1}a = \sigma^{-1}(a)(u_\sigma)^{-1}$

iii) $(bu_\sigma)^{-1} = \sigma^{-1}(b^{-1})(u_\sigma)^{-1}$

can be used to verify the chain of equalities

$$\begin{aligned}
\psi[(au_\sigma)\cdot(bu_\tau)] &= \psi[b\ \tau(a)\ g(\tau,\sigma)u_{\tau\sigma}] \\
&= \sigma^{-1}\tau^{-1}\ (b\tau(a)\ g(\tau,\sigma))\ (u_{\tau\sigma})^{-1} \\
&= \sigma^{-1}\tau^{-1}\ (b)\sigma^{-1}(a)\ (u_\tau u_\sigma)^{-1} \\
&= \sigma^{-1}(a)(u_\sigma)^{-1}\tau^{-1}(b)(u_\tau)^{-1} \\
&= \psi(au_\sigma)\psi(bu_\tau)
\end{aligned}$$

so ψ is an R-algebra isomorphism.

Let η: $\Delta(S:G:g) \to S \otimes S$ be defined by $\eta(\Sigma_{\tau \in G} s_\tau u_\tau) = \Sigma_{\tau \in G}(s_\tau \otimes 1)e_{\tau^{-1}}$ where the e_τ are the orthogonal idempotents in $S \otimes S$ defined in 2.12. It is clear that η is an R-module isomorphism. It is also clear that $\Delta(S:G:f) \otimes \Delta(S:G:g) \simeq \Delta(S \otimes S:G \times G:(f,g))$ under the correspondence $(s_\sigma u_\sigma \otimes s_\tau u_\tau) \to (s_\sigma \otimes s_\tau)u_{(\sigma,\tau)}$ so there is a chain of R-algebra isomorphisms

$$\Delta(S \otimes S:\ G \times G:(g,g^{-1})) \to \Delta(S:G:g) \otimes \Delta(S:G:g^{-1})$$

$$\overset{1\otimes\psi^{-1}}{\to} \Delta(S:G:g) \otimes \Delta(S:G:g)^\circ$$

$$\overset{\phi}{\to} \mathrm{Hom}_R(\Delta(S:G:g),\ \Delta(S:G:g))$$

$$\overset{\bar{\eta}}{\to} \mathrm{Hom}_R(S \otimes S,\ S \otimes S)$$

$$\overset{j^{-1}}{\to} \Delta(S \otimes S:\ G \times G:1).$$

This chain of maps is given explicitly by

$$(s_\sigma \otimes s_\tau)\ u_{(\sigma,\tau)} \to s_\sigma u_\sigma\ s_\tau u_\tau$$

$$\to \delta = s_\sigma u_\sigma \otimes \tau^{-1}(s_\tau)\ g(\tau^{-1},\tau)^{-1}u_{\tau^{-1}}$$

$$\to \phi(\delta)\ \text{(left multiplication by } \delta)$$

$$\to \eta\cdot\phi(\delta)\cdot\eta^{-1}$$

$$\to j^{-1}(\eta\cdot\phi(\delta)\cdot\eta^{-1})$$ (j is given in Proposition 3.1.2).

Take $z \otimes w \in S \otimes S$, then by a straight forward computation we have

$$\eta\phi(\delta)\eta^{-1}(z\otimes w) = j\Sigma_{\rho \in G}(g(\sigma,\rho)g(\sigma\rho,\tau^{-1}) \otimes \tau g(\tau^{-1},\tau)^{-1}e_{\tau\rho^{-1}\sigma^{-1}})$$
$$s_\sigma \otimes s_\tau\cdot u_{(\sigma,\tau)}(z\otimes w).$$

So letting $h(\sigma,\tau) = \Sigma_{\rho \in G} g(\sigma,\rho)g(\sigma\rho,\tau^{-1}) \otimes \tau g(\tau^{-1},\tau)^{-1}e_{\tau\rho^{-1}\sigma^{-1}}$

for all $(\sigma,\tau) \in G \times G$ we have shown the correspondence $s_\sigma \otimes s_\tau\cdot u_{(\sigma,\tau)} \to s_\sigma \otimes s_\tau\ h(\sigma,\tau)\ u_{(\sigma,\tau)}$ is an R-algebra isomorphism from $\Delta(S \otimes S:G \times G:(g,g^{-1}))$ to $\Delta(S \otimes S:G \times G:1)$. One can check that this implies $(g,g^{-1}) \in B^2(G \times G, U(S \otimes S))$.

To complete the proof that α_4 is a homomorphism observe that for any $f,g \in Z^2(G,U(S))$ there is a natural chain of algebra isomorphisms

$$\begin{aligned}\Delta(S:G:f) \otimes \Delta(S:G:g) &\to \Delta(S \otimes S:G \times G:(f,g)) \\ &\to \Delta(S \otimes S:G \times G:(fg,1)) \\ &\to \Delta(S:G:fg) \otimes \Delta(S:G:1).\end{aligned}$$

Since $\Delta(S:G:1) = \mathrm{Hom}_R\ (S,S)$ we have

$$\Delta(S:G:f) \otimes \Delta(S:G:g) = \Delta(S:G:fg) \text{ in } B(R).$$

2.14 <u>Exactness at $H^2(G,U(S))$</u>: Let $|V| \in P(S)^G$ and assume $\alpha_3(|V|) = [f]$. Then for each $\sigma \in G$ there is an isomorphism $\psi_\sigma : V \to V$ so that $\psi_\sigma(sx) = \sigma(s)\psi_\sigma(x)$ for all $s \in S$ and $x \in V$. The equations

$$f(\sigma,\tau) = \psi_{\sigma\tau}\psi_\tau^{-1}\psi_\sigma^{-1} \text{ for all } \sigma,\tau \in G \text{ define } f.$$

Observe that V can be made into a $\Delta(S:G:f^{-1})$-module by defining $u_\sigma \cdot x = \psi_\sigma(x)$ for all $\sigma \in G$, $x \in V$. Thus there is an R-algebra homomorphism ϕ given by left multiplication from $\Delta(S:G:f^{-1})$ to $\mathrm{Hom}_R\ (V,V)$. Now $\Delta(S:G:f^{-1})$ is a central separable R-algebra and V is a faithful R-module so V is a faithful $\Delta(S:G:f^{-1})$-module and therefore ϕ is a monomorphism. If Γ is the centralizer of $\Delta(S:G:f^{-1})$ in $\mathrm{Hom}_R\ (V,V)$ then $\mathrm{Hom}_R\ (V,V) = \Gamma \otimes \Delta(S:G:f^{-1})$ by Theorem 2.4.3. Observe that both $\mathrm{Hom}_R\ (V,V)$ and $\Delta(S:G:f^{-1})$ have the same rank over R so $\mathrm{Rank}_R\ (\Gamma) = 1$ and ϕ is an isomorphism. We have shown image $\alpha_3 \subseteq$ kernel α_4.

On the other hand let $[f] \in$ kernel α_4, then since α_4 is a group homomorphism $[f^{-1}] \in$ kernel α_4. Therefore there is a finitely generated projective faithful R-module V with $\Delta(S:G:f^{-1}) \simeq \mathrm{Hom}_R\ (V,V)$. We can view V as an S-module since $S \cdot u_1 \subseteq \Delta(S:G:f^{-1})$. Moreover $\mathrm{Hom}_S\ (V,V)$ is just the centralizer of S in $\mathrm{Hom}_R(V,V) = \Delta(S:G:f^{-1})$ so $\mathrm{Hom}_S(V,V) = S$. Thus $|V| \in P(S)$ and each of the elements $u_\sigma \in \Delta(S:G:f^{-1})$ gives an

R-isomorphism $\psi_\sigma : V \to V$ by $\psi_\sigma(x) = u_\sigma x$ for all $x \in V$. Now $\psi_\sigma(sx) = \sigma(s)\psi_\sigma(x)$ for all $s \in S$, $x \in V$ so $V \simeq \sigma \cdot V$ as S-modules and $|V| \in P(S)^G$. Also observe that $\psi_{\sigma\tau}\psi_\tau^{-1}\psi_\sigma^{-1} = u_{\sigma\tau}(u_\tau)^{-1}(u_\sigma)^{-1} = f(\sigma,\tau)$ so $\alpha_3(|V|) = [f]$ and kernel α_4 = image α_3 .

2.15 <u>The definition of α_5:</u> To facilitate the definition of α_5 we will make P(S) into a G-module in a formally different way than before. If $|V| \in P(S)$ and $\sigma \in G$ we let $\sigma \cdot |V| = |{}_\sigma V|$ where ${}_\sigma V$ is the R-module V with action from S by $s \cdot x = \sigma^{-1}(s)x$ for all $s \in S$, $x \in V$. Let $[A] \in B(S/R)$, then by Theorem 2.5.5 there is a central separable R-algebra equivalent to A in B(R) and containing S as a maximal commutative subalgebra. We assume S is a subalgebra of A and $A^S = S$. Since $A^S = S$ we know $S \otimes A^o = \mathrm{Hom}_S(A,A)$ and A is a finitely generated projective faithful S-module. We are in the context of Proposition 1.3.3 so every left $S \otimes A^o$-module is of the form $X \otimes {}_S A$ for a uniquely determined S-module X. If Y is a left $S \otimes A^o$-module let ${}_\sigma Y$ be the $S \otimes A^o$-module equal to Y as an $R \otimes A^o$-module and with $(s \otimes a) \cdot y = \sigma^{-1}(s)ya$ for all $s \in S$, $a \in A$, $y \in Y$. Then there is a unique S-module ${}^\sigma M$ with

$${}_\sigma A = {}^\sigma M \otimes_S A \text{ as } S \otimes A^o\text{-modules.}$$

A count of ranks shows $|{}^\sigma M| \in P(S)$. Also one can check the following chain of natural $S \otimes A^o$-module isomorphisms.

$$\begin{aligned} {}^{\sigma\tau}M \otimes_S A &\simeq {}_{\sigma\tau}A \\ &\simeq {}_\sigma({}_\tau A) \\ &\simeq {}_\sigma({}^\tau M \otimes_S A) \\ &\simeq {}^\tau_\sigma M \otimes_S ({}_\sigma A) \\ &\simeq ({}^\tau_\sigma M \otimes_S {}^\sigma M) \otimes_S A . \end{aligned}$$

Define $f : G \to P(S)$ by $f(\sigma) = |{}^\sigma M|$. We have shown that $f(\sigma\tau) = f(\sigma)\sigma \cdot f(\tau)$

so $f \in Z^1(G,P(S))$. The correspondence α_5 is defined by $\alpha_5([A]) = [f]$.

2.16 α_5 is well defined: Keeping the notation of the preceding section let $D \in B(S/R)$ with D containing S as a maximal commutative subalgebra and $[D] = [A]$ in $B(R)$. There are finitely generated projective faithful R-modules P_1 and P_2 with $B \cong A \otimes \mathrm{Hom}_R(P_1,P_1) \simeq D \otimes \mathrm{Hom}_R(P_2,P_2)$. If $P_i^* = \mathrm{Hom}_R(P_i,R)$ then the modules $A \otimes P_1^*$ and $D \otimes P_2^*$ are left $S \otimes B^o$-modules with

$$\begin{aligned}\mathrm{Hom}_S(A \otimes P_1^*, A \otimes P_1^*) &\simeq \mathrm{Hom}_S(A,A) \otimes \mathrm{Hom}_R(P_1^*,P_1^*)\\ &\simeq S \otimes A^o \otimes \mathrm{Hom}_R(P_1^*,P_1^*)\\ &\simeq S \otimes B^o\end{aligned}$$

and similarly

$$\mathrm{Hom}_S(D \otimes P_2^*, D \otimes P_2^*) \simeq S \otimes D^o \otimes \mathrm{Hom}_R(P_2^*,P_2^*) \simeq S \otimes B^o.$$

By Theorem 2.6.4 there is an element $J \in P(S)$ so that $(A \otimes P_1^*) \otimes_S J \simeq D \otimes P_2^*$ as $S \otimes B^o$-modules.

Let ${}^{\sigma}M \otimes_S A \cong {}_{\sigma}A$ as $S \otimes A^o$-modules and let $W^{\sigma} \otimes_S D \cong {}_{\sigma}D$ as $S \otimes D^o$-modules for each $\sigma \in G$. Let $f,g{:}G \to P(S)$ be defined by

$$f(\sigma) = |{}^{\sigma}M| \text{ for each } \sigma \in G$$

$$g(\sigma) = |{}^{\sigma}W| \text{ for each } \sigma \in G.$$

Now

$$\begin{aligned}{}_{\sigma}(D \otimes P_2^*) &\cong {}_{\sigma}(A \otimes P_1^* \otimes_S J)\\ &\cong {}_{\sigma}J \otimes_S ({}_{\sigma}A) \otimes P_1^*\\ &\cong ({}_{\sigma}J \otimes_S {}^{\sigma}M \otimes_S A) \otimes P_1^*\end{aligned}$$

and

$$\begin{aligned}{}_{\sigma}(D \otimes P_2^*) &\cong {}_{\sigma}D \otimes P_2^*\\ &\cong {}^{\sigma}W \otimes_S D \otimes P_2^*.\end{aligned}$$

Recall that $|J^*| = |\mathrm{Hom}_S(J,S)|$ is the inverse of $|J|$ in $P(S)$ so

$$({}_\sigma J \otimes_S J^* \otimes_S {}^\sigma M) \otimes_S (J \otimes_S A \otimes P_1^*) \simeq {}^\sigma W \otimes_S D \otimes P_2^*$$
$$\simeq {}^\sigma W \otimes_S (J \otimes_S A \otimes P_1^*).$$

By Corollary 1.3.4

$${}_\sigma J \otimes_S J^* \otimes_S {}^\sigma M \cong {}^\sigma W \text{ for each } \sigma \in G.$$

Define $b: G \to P(S)$ by $b(\sigma) = |{}_\sigma J \otimes_S J^*|$, then $b \in B^1(G, P(S))$ and $b \cdot f = g$ so $[f] = [g]$ in $H^1(G, P(S))$ and α_5 is well defined.

2.17 α_5 is a homomorphism: Let $[A_1]$ and $[A_2]$ be elements in $B(S/R)$ and assume S is a maximal commutative subalgebra of A_i $(i = 1,2)$. Let $B = A_1 \otimes A_2$, then $S \otimes S$ is a maximal commutative R-subalgebra of B. Let e be a separability idempotent for S in $S \otimes S$. Observe that $\mathrm{Hom}_B(Be,Be) = (eBe)^o$ since every element in $\mathrm{Hom}_B(Be,Be)$ is a right multiplication by an element in eBe. Thus by Theorem 2.4.3 we know $\mathrm{Hom}_R(Be,Be) = B \otimes (eBe)^o$ and $[B] = [eBe]$ in $B(R)$. $(S \otimes S)e = (S \otimes 1)e = S$ is a maximal commutative subalgebra of eBe for if $z \in (eBe)^S$ then $z \in eBe \cap B^{S\otimes S} \subseteq (S \otimes S)e = S$. Following the procedure outlined in 2.15 let

$${}_\sigma(eBe) = {}^\sigma M \otimes_S eBe \text{ for all } \sigma \in G$$

and let $f: G \to P(S)$ be defined by $f(\sigma) = |{}^\sigma M|$, then $[f] = \alpha_5[eBe] = \alpha_5[A_1 \otimes A_2]$.

For $i = 1,2$ let $f_i: G \to P(S)$ be defined by $f_i(\sigma) = |{}_i^\sigma W|$ where ${}_\sigma A_i = {}_i^\sigma W \otimes_S A_i$. Then $\alpha_5[A_i] = [f_i]$. We want to show that $[f] = [f_1][f_2]$. Let $\sigma \in G$ and let σ act on $S \otimes S$ by $\sigma(x \otimes y) = \sigma(x) \otimes \sigma(y)$ for all $x, y \in S$. The module $Be = (A_1 \otimes A_2)e$ so

$${}_\sigma[(A_1 \otimes A_2)e] = ({}_\sigma A_1 \otimes {}_\sigma A_2)e.$$

Since $e \otimes e \in (S \otimes S) \otimes (eBe)^o$ we see that for any $(S \otimes S) \otimes (eBe)^o$-module V, the module $(e \otimes 1)V$ is an $(S \otimes S)e \otimes (eBe)^o = S \otimes (eBe)^o$ -

module. Because $(s \otimes 1 - 1 \otimes s)e = 0$ for all $s \in S$ and $\sigma(e) = e$ for all $\sigma \in G$ one can check that

$$e[{}_{\sigma}(A_1 \otimes A_2)e] \simeq {}_{\sigma}(eBe).$$

Therefore

$$e[{}_{\sigma}(A_1 \otimes A_2)e] \simeq e[({}_1^{\sigma}W) \otimes_S ({}_2^{\sigma}W) \otimes_S A_1 \otimes A_2]e$$

$$\simeq ({}_1^{\sigma}W) \otimes_S ({}_2^{\sigma}W) \otimes_S e(A_1 \otimes A_2)e.$$

Thus $({}_1^{\sigma}W) \otimes_S ({}_2^{\sigma}W) \otimes_S eBe = {}^{\sigma}M \otimes_S eBe$ so by Proposition 1.3.3 we have $({}_1^{\sigma}W) \otimes_S ({}_2^{\sigma}W) \simeq {}^{\sigma}M$. Therefore $[f] = [f_1]\,[f_2]$ and α_5 is a homomorphism.

2.18 Exactness at $B(S/R)$: Let $[A] \in B(S/R)$ with S a maximal commutative subalgebra of A and $\alpha_5[A] = [1]$ in $H^1(G,P(S))$. Keeping the notation and terminology of sections 2.11-2.17 the hypothesis on A implies there is a $|J| \in P(S)$ with ${}_{\sigma}A = \sigma \cdot J \otimes_S J^* \otimes_S A$ as $S \otimes A^o$-modules.

Observe that $\sigma \cdot J \otimes_S J^* \simeq \sigma \cdot S$ as S-modules so ${}_{\sigma}A \simeq \sigma \cdot S \otimes_S A$. Let θ_{σ} be the $S \otimes A^o$-isomorphism from ${}_{\sigma}A$ to $\sigma \cdot S \otimes_S A$. Let

$$\theta_{\sigma}(1) = \sum_{i=1}^{n} s_i \otimes a_i$$

$$= \sum_{i=1}^{n} 1 \otimes \sigma^{-1}(s_i)a_i$$

$$= 1 \otimes u_{\sigma}.$$

Also in ${}_{\sigma}A$ we know that for all $s \otimes a \in S \otimes A^o$

$$(s \otimes a)\cdot 1 = \sigma^{-1}(s)\, a = (1 \otimes \sigma^{-1}(s)a)\cdot 1.$$

Applying θ_{σ} to this equation we have

$1 \otimes su_{\sigma}a = (s \otimes a)\cdot\theta_{\sigma}(1) = (1 \otimes \sigma^{-1}(s)a)\cdot\theta_{\sigma}(1) = 1 \otimes u_{\sigma}\sigma^{-1}(s)\, a$ for all $s \in S$, $a \in A$, $\sigma \in G$. Since θ_{σ} is an isomorphism, $u_{\sigma} \in U(A)$. Letting $a = 1$ in the equations above we find that $\sigma(s)u_{\sigma} = u_{\sigma}s$ for all $s \in S$, $\sigma \in G$, and $\sigma(s) = u_{\sigma}su_{\sigma}^{-1}$ for all $s \in S$. Let $f: G \times G \to U(S)$ be defined by $f(\sigma,\tau) = u_{\sigma}u_{\tau}u_{\sigma\tau}^{-1}$ for all σ, $\tau \in G$. Observe that f is well defined since $f(\sigma,\tau)$ is an isomorphism in $\mathrm{Hom}_S(S,S)$. Moreover $u_\sigma u_\tau = f(\sigma,\tau)u_{\sigma\tau}$ so the subalgebra $\sum_{\sigma\in G} Su_\sigma$ of A is a homomorphic image of $\Delta(S:G:f)$. Since $R \subseteq S$ the natural homomorphism from $\Delta(S:G:f)$ onto $\sum_{\sigma\in G} Su_\sigma$ is an isomorphism. An application of Theorem 2.4.5 together with the hypothesis that S is a maximal commutative subalgebra of A shows $A = \sum_{\sigma\in G} Su_\sigma = \Delta(S:G:f)$ proving that kernel $\alpha_5 \subseteq$ image α_4.

Finally let $[f] \in H^2(G,U(S))$ and let $A = \Delta(S:G:f)$. For each basis element $u_\sigma \in \Delta(S:G:f)$ define an isomorphism $\theta_\sigma: {}_\sigma A \to \sigma{\cdot}S \otimes_S A$ by $\theta_\sigma(1) = 1 \otimes u_\sigma$. One can check that θ_σ is an $S \otimes A^o$-isomorphism which proves that if $[g] = \alpha_5[A]$ then $g(\sigma) = |\sigma(S)|$ for all $\sigma \in G$. Thus $g \in B^1(G,P(S))$ and image $\alpha_4 \subseteq$ kernel α_5 completing the proof.

Exercise (Kummer theory)

Let R be a commutative ring with no idempotents other than 0 and 1. Let n be a positive integer with $n\cdot 1$ a unit in R and R containing a primitive n^{th} root of 1. Assume the class group of R is trivial. Let $a_1,\ldots,a_n$ be units in R, a splitting ring for the separable polynomial $\prod_{i=1}^{r} (x^n - a_i)$ in $R[x]$ is called a Kummer extension of R. Prove that N is a Kummer extension of R if and only if N is a normal separable extension of R with no idempotents other than 0 and 1, the Galois group G of N over R is abelian, and the exponent of G is a divisor of n. (See [A] p. 61).

CHAPTER V

In this chapter results on the structure of central separable algebras and on the Brauer group of Dedekind domains are presented. We then conclude with some historical remarks, suggestions of problems for future study, and applications to various areas in ring theory.

§ 1. Structure Theory

Assume throughout this section that R is a commutative ring and that A and B are central separable R-algebras equivalent in B(R). Then $A \otimes B^o \simeq \mathrm{Hom}_R(M,M)$ and $B \otimes B^o \simeq \mathrm{Hom}_R(N,N)$ for some finitely generated projective faithful R-modules M and N.

Theorem 2.4.3 implies $A \simeq \mathrm{Hom}_{B^o}(M,M)$, $B^o \simeq \mathrm{Hom}_B(N,N)$ where M is a B^o-progenerator and N is a B-progenerator. It follows from Proposition 1.3.3 that the categories ${}_A\mathfrak{M}$ and ${}_B\mathfrak{M}$ are naturally equivalent. This, for example, implies the isomorphism classes of indecomposable projective A-modules are in one-to-one correspondence with the isomorphism classes of indecomposable projective B-modules. For questions whose answers are invariant under the equivalence of Proposition 1.3.3 we are free to choose a convenient representative central separable algebra from the given class in B(R).

Now assume the only idempotents in R are 0 and 1, and let P(R) be the class group of R (See Chap. I Sect. 5). We will call two indecomposable finitely generated projective A-modules M and N equivalent in case there is an element $|X| \in P(R)$ with $M \simeq N \otimes X$ as A-modules. One easily checks that this relation defines an equivalence relation on the set of isomorphism classes of indecomposable finitely generated projective A-modules.

Theorem 1.1: Let A be a central separable algebra over the commutative ring R. Assume R has no idempotents other than 0 and 1. Then there is a natural one-to-one correspondence between the equivalence classes of indecomposable finitely generated projective A-modules and the isomorphism classes of central separable R-algebras with no idempotents other than 0 and 1 equivalent to A in B(R). If M is an indecomposable finitely generated projective A-module the class containing M corresponds to $\mathrm{Hom}_A(M,M)^o$.

Proof: Let M be an indecomposable finitely generated projective A-module and let $B = \mathrm{Hom}_A(M,M)^o$. Then by Theorem 2.4.3 we have $A \otimes B^o \simeq \mathrm{Hom}_R(M,M)$ and B is a central separable R-algebra which is equivalent to A in B(R). An idempotent in B defines an A-projection on M so since M is indecomposable the only idempotents in B are 0 and 1. If $|X| \in P(R)$ then $M \otimes X$ is indecomposable since a decomposition of $M \otimes X$ would imply a decomposition of $M \otimes X \otimes X^* \simeq M$. There is a natural isomorphism (1.2.4) from $\mathrm{Hom}_A(M,M) \otimes \mathrm{Hom}_R(X,X)$ to $\mathrm{Hom}_A(M \otimes X, M \otimes X)$. Since $\mathrm{Hom}_R(X,X) \simeq R$, we have $\mathrm{Hom}_A(M \otimes X,\ M \otimes X) \simeq B^o$.

If B is a central separable R-algebra with no idempotents other than 0 and 1 and equivalent to A in B(R) then there is a finitely generated projective R-module M with $A \otimes B^o \simeq \mathrm{Hom}_R(M,M)$. Thus M is a finitely generated projective A-module and $B \simeq \mathrm{Hom}_A(M,M)^o$. Since an A-projection on M corresponds to an idempotent in B, M is an indecomposable A-module.

Finally, let N be another finitely generated projective indecomposable A-module with $B \simeq \mathrm{Hom}_A(N,N)^o$. Then $A \otimes B^o \simeq \mathrm{Hom}_R(N,N) \simeq \mathrm{Hom}_R(M,M)$. By Lemma 2.6.4 there is an $|X| \in P(R)$ so that $M \simeq N \otimes X$ as $A \otimes B^o$-modules which proves the theorem.

An algebra A over a field R is called a <u>central simple</u> R-algebra in case the only two-sided ideals in A are 0 and A, the center of A is R, and A is finite dimensional as a vector space over R.

<u>Proposition 1.2</u>: Let R be a field and let A be an R-algebra. Then A is a central simple R-algebra if and only if A is a central separable R-algebra.

<u>Proof</u>: If A is a central separable R-algebra then by Proposition 2.2.1 and Lemma 2.3.5 A is a central simple R-algebra. For the converse we show $K \otimes A$ is a central simple K-algebra for every extension field K of R and apply Theorem 2.2.5.

Let K be an extension field of R and let $0 \neq I$ be a two-sided ideal in $K \otimes A$. We need only show I contains an element of the form $k \otimes a \neq 0$ for $k \in K$, $a \in A$. For then I contains $K \otimes a$ so contains $K \otimes AaA = K \otimes A$. Suppose by way of contradiction that I does not contain any non-zero elements of the form $k \otimes a$. Among all elements

$$b = \Sigma_{i=1}^{s} k_i \otimes a_i \in I \qquad k_i \in K,\ a_i \in A$$

with the k_i linearly independent over R choose a non-zero element $b \in I$ for which s is as small as possible. Then $s > 1$ and $a_i \neq 0$. Keeping $k_1, \ldots, k_s$ fixed, consider the set A_1 of all elements $a_1' \in A$ such that

$$k_1 \otimes a_1' + \Sigma_{i=2}^{s} k_i \otimes a_i' \in I$$

for some elements $\{a_i'\} \subseteq A$. One quickly checks that A_1 is a non-zero ideal in A so since A is simple we know $A_1 = A$. Therefore

$$b^* = k_1 \otimes 1 + \Sigma_{i=2}^{s} k_i \otimes a_i' \in I \text{ for some}$$

$\{a_i'\} \subseteq A$. Then, for all $a \in A$,

$$b^* a - ab^* \in I$$

and it follows that

$$\Sigma_{i=2}^{s} k_i \otimes (a_i'a - aa_i') \in I.$$

By minimality of s, this element is zero, and since the $\{k_i\}$ are linearly independent over R we have

$$a_i'a = aa_i' \qquad i = 2, \ldots, s,\ a \in A .$$

Therefore, $\{a_i'\} \subseteq R$ and since we are tensoring with respect to R we have

$$b^* = (k_1 + \Sigma_{i=2}^{s} a_i' k_i \otimes 1) \in I .$$

This contradicts the choice of s and proves the only two-sided ideals in $K \otimes A$ are $\{0\}$ and $K \otimes A$, which proves the proposition.

If A is an algebra over a field R then an A-module M is called simple in case M has no submodules except $\{0\}$ and M .

Corollary 1.3: Let A be a central simple (= central separable) algebra over the field R, then

1. Every left A-module is projective.
2. Every left ideal in A is generated by an idempotent.
3. Any indecomposable left A-module is simple.
4. Any two simple left A-modules are isomorphic.
5. There is a unique (up to isomorphism) central division algebra D over R equivalent to A in B(R), and A is the ring of all $n \times n$ matrices with entries from D.

Proof: Proposition 2.2.3 implies that all A-modules are projective. Thus if N is a submodule of the A-module M the sequence $0 \to N \to M \to M/N \to 0$ splits. Appropriate choices for M and N yield statements 2 and 3.

Since A is a finite dimensional vector space over R we can find a simple left ideal I in A which by 2. is generated by an idempotent e. If M is any simple left A-module then since A is simple M is a generator module and we can find an A-homomorphism f mapping M to $I = Ae$. By Schur's lemma, f is an isomorphism which proves 4. To prove 5 apply Theorem 1.1 to get a unique central separable algebra D with no idempotents but 0 and 1, equivalent to A in B(R) and with $A \simeq \mathrm{Hom}_D(M,M)^o$ for a simple left A-module M. By 2., any left ideal in D contains an idempotent so D has no proper left ideals except $\{0\}$ and therefore is a division algebra. To complete the proof observe that $\mathrm{Hom}_D(M,M)^o$ is isomorphic to $n \times n$ matrices with entries in D where n is the dimension of M over D.

We conclude this section by demonstrating the existence of a separable splitting field for any central simple algebra.

<u>Lemma 1.4</u>: Let R be a field and let $R(\alpha)$ be an algebraic field extension of R. Then

1. If there is a non-identity R-automorphism of $R(\alpha)$, then $R(\alpha)$ contains a separable extension of R not equal to R.

2. If $R(\alpha)$ contains no separable extension of R not equal to R then the characteristic of R is a prime p and $\alpha^{p^n} \in R$ for some non-negative integer n.

<u>Proof</u>: We refer the reader to S. Lang's book, <u>Algebra</u>, Chapter VII.

<u>Lemma 1.5</u>: Let D be a central division algebra over the field R with $D \neq R$. Then D contains a subfield which is separable over R and properly contains R.

<u>Proof</u>: Let $\alpha \in D$, $\alpha \notin R$, then α is algebraic over R since D is finite dimensional. Thus $R(\alpha)$ is an algebraic field extension of R and by Lemma 1.4 we can assume that there is an integer n with $u = \alpha^{p^n} \notin R$ but $\alpha^{p^{n+1}} = u^p \in R$. Let σ be the R-automorphism of D given by $\sigma(x) = u\, x\, u^{-1}$. Then $\sigma^p = 1$ but $\sigma \neq 1$ so $(\sigma-1) \neq 0$ but $(\sigma-1)^p = 0$ in $\mathrm{Hom}_R(D,D)$. Let r be the largest integer such that $(\sigma-1)^r \neq 0$, and let $y \in D$ such that $(\sigma-1)^r(y) \neq 0$. Suppose $a = (\sigma-1)^{r-1}(y)$ and $b = (\sigma-1)(a) \neq 0$. By definition of r, $\sigma(b) = b$, and if $c = b^{-1}a$, $\sigma(c) = b^{-1}\sigma(a) = b^{-1}(b+a) = 1 + c \neq c$. Now the restriction of σ to $R(c)$ is an automorphism of $R(c)$ distinct from 1 so by Lemma 1.4 $R(c)$ contains a separable extension of R not equal to R which proves the lemma.

<u>Theorem 1.6</u>: Let A be a central simple algebra over the field R. Then there is a normal separable field extension N of R which is a splitting field for A.

Proof: By Corollary 1.3 A is equivalent to a central division algebra D in B(R). If N splits D, then N will split A. Let L be a maximal separable subfield of D over R. By Theorem 2.4.3, D^L is a central division algebra over L. If $D^L \neq L$ then D^L contains a separable extension L' of L properly containing L by Lemma 1.5. But L' is separable over R and thus contradicts the maximality of L so $D^L = L$. Thus by Theorem 2.5.4, L is a splitting field for D. By Theorem 3.2.8 we can imbed L in a normal separable extension N of R with no idempotents other than 0 and 1. It follows (Exercise #6 of Chapter III) that N is a field. Now $N \otimes_R D \cong N \otimes_L L \otimes_R D$ so N is a splitting field for D.

§ 2. The Brauer Group of a Dedekind Domain

In Chapter 1, Section 5 a Dedekind domain was defined to be a noetherian integral domain over which every fractional ideal is projective. We next list without proof additional facts we will employ concerning Dedekind domains and refer the reader to Zariski and Samuel [L] or Curtis and Reiner [G] for the proofs.

Lemma 2.1: Let R be a Dedekind domain with quotient field K.

1. If $\sigma \in K$ satisfies a monic polynomial with coefficients in R, then $\sigma \in R$.

2. If $\mathfrak{p}$ is a prime ideal in R then $R_{\mathfrak{p}}$ is a local principal ideal domain.

3. If M is a finitely generated, torsion-free R-module, then M is projective.

An integral domain satisfying 1 of Lemma 2.1 is said to be integrally closed in its quotient field. The natural monomorphism from the integral domain R to its quotient field K induces a homomorphism from B(R) to B(K).

Lemma 2.2: Let R be a Dedekind domain with quotient field K, then the natural homomorphism from B(R) to B(K) is a monomorphism.

<u>Proof</u>: Let [A] ϵ B(R) and assume K $\otimes$ A $\simeq$ $Hom_K(V,V)$ for some finite dimensional vector space V over K. View each element $\sigma \epsilon$ K $\otimes$ A as a linear transformation on V, let $0 \neq v \epsilon V$, and define the K $\otimes$ A epimorphism ψ: K $\otimes$ A $\rightarrow$ V by $\psi(\sigma) = \sigma(v)$ for all $\sigma \epsilon$ K $\otimes$ A.

Observe that E = ψ(A) is a finitely generated, torsion-free R-submodule of V so by Lemma 2.1 E is a projective R-module. (Also $Hom_R(E,E)$ is a central separable R-subalgebra of K $\otimes$ A). Since E is an A-module which is faithful as an R-module Corollary 2.3.1 implies E is a faithful A-module so we can view A as an R-subalgebra of $Hom_R(E,E)$. By employing Theorem 2.4.3 and counting ranks we know the centralizer of A in $Hom_R(E,E)$ is R and therefore A $\simeq$ $Hom_R(E,E)$ which proves the lemma.

<u>Lemma 2.3</u>: Let A be a central separable algebra over the local ring R. Then every finitely generated left ideal in A is principal.

<u>Proof</u>: Let N = Rad(R), then N·A = Rad(A) and A/NA is a central simple algebra over the field R/N. If I is a finitely generated left ideal in A, then by Corollary 1.3, I/NI is generated by a single element x + NI with x ϵ I. Thus Ax + NI = I. An easy consequence of the finite generation of I and the properties of N is that Ax = I.

We now fix notation for the rest of this section. Let R be a Dedekind domain with quotient field K and let Σ be a central simple K-algebra. If an element a ϵ Σ is contained in a finitely generated R-submodule of Σ then R(a) is finitely generated over R since R is noetherian. Observe that R(a) is finitely generated over R if and only if a satisfies a monic polynomial with coefficients in R. An R-subalgebra A of Σ is called an <u>order</u> over R in Σ in case A contains a K-basis for Σ and every element in A is contained in a finitely generated R-submodule of Σ. The order A over R in Σ is called <u>maximal</u> in case A is not properly contained in any other order over R in Σ. It is clear that the union of a chain of R-orders in Σ is again an R-order so every R-order is contained in a maximal order. Let $u_1,\dots,u_n$ be a K-basis for Σ with

$$u_i u_j = \Sigma_{k=1}^{n} \sigma_{ij}^{k} u_k \qquad \sigma_{ij}^{k} \in K .$$

Then $\sigma_{ij}^k = \beta_{ij}^k/\gamma_{ij}^k$ with $\beta_{ij}^k, \gamma_{ij}^k \in R$, $\gamma_{ij}^k \neq 0$. Let $d = \Pi_{i,j,k}\, \gamma_{ij}^k$, then $d \in R$ and $du_1,\dots,du_n$ is a K-basis for Σ. Moreover, $du_1,\dots,du_n$ form an R-basis for a finitely generated R-subalgebra of Σ. Since R is noetherian this R-subalgebra is an order over R in Σ. We have therefore proved

Lemma 2.4: Let R be a Dedekind domain with quotient field K, and let Σ be a central simple K-algebra. Then there is an order over R in Σ. Any order over R in Σ is contained in a maximal order so maximal orders over R in Σ exist.

Let N be a normal separable splitting field for Σ. Then $N \otimes_K \Sigma \simeq M_n(N)$ (n x n matrices with entries from N for some positive integer n). Let h be the isomorphism. Then for any $x \in \Sigma$ we can view x as the matrix $h(1 \otimes x)$ and let $rt(x) = \text{trace}[h(1 \otimes x)]$. We will show

1. rt is independent of the choice of h
2. $rt \in \text{Hom}_K(\Sigma, K)$
3. rt is independent of the choice of N.

If $g = N \otimes_K \Sigma \to M_n(N)$ is another isomorphism then $h^{-1}g$ is an automorphism of $N \otimes_K \Sigma$ which is inner by Proposition 2.6.3 . Thus for any $x \in A$, $g(x) = uh(x)u^{-1}$ for a fixed $u \in M_n(N)$, so trace $[g(x)] = \text{trace}[h(x)]$. If σ is an automorphism of N leaving K fixed, then $\sigma \otimes 1$ is a K-automorphism of $N \otimes_K \Sigma$, and σ can also be viewed as a K-automorphism of $M_n(N)$. Then $\sigma \cdot h \cdot \sigma^{-1} \otimes 1$ is an N-isomorphism g from $N \otimes_K \Sigma$ to $M_n(N)$ and for each $x \in \Sigma$, $\sigma \cdot h \cdot \sigma^{-1} \otimes 1\ (1 \otimes x) = \sigma h(1 \otimes x)$ so by the preceding paragraph, $\text{trace}[\sigma \cdot h(1 \otimes x)] = \text{trace}[h(1 \otimes x)]$. Also, $\text{trace}[\sigma h(1 \otimes x)] = \sigma[\text{trace}\ h(1 \otimes x)]$. Since exactly K is left fixed by the K-automorphisms of N, $\text{trace}[h(1 \otimes x)] \in K$.

If N' is another normal separable splitting field for Σ then there is a normal separable field extension E of K containing N and N'. Then $E \otimes_K \Sigma = E \otimes_N N \otimes_K \Sigma$ so there is a natural extension of h to an isomorphism $\overline{h}$ from $E \otimes_K \Sigma$ to $M_n(E)$. But $\overline{h}(1 \otimes x) = h(1 \otimes x)$. Apply the same reasoning to N' and use part 1 to conclude that rt is independent of the choice of N.

The linear transformation rt from Σ to K which we have defined is called the <u>reduced trace</u>. If A is an order over R in Σ and $b \in A$ then $R[b]$ is a finitely generated R-module so $R[rt(b)] \subseteq rt(R[b])$ is a finitely generated R-module. Since $rt(b) \in K$, Lemma 2.1 implies $rt(b) \in R$. Thus $rt(A) \subseteq R$.

<u>Lemma 2.5</u>: Let Σ be a central simple K-algebra. Then the natural K-homomorphism ψ from Σ to $Hom_K(\Sigma,K)$ given by $\psi(a) = rt(a-)$ is an isomorphism.

<u>Proof:</u> Let N be a normal separable splitting field for Σ, then rt is the trace of the elements in $N \otimes_K \Sigma$ when viewed as matrices over N. In $N \otimes_K \Sigma$ we can find elements x,y so that xay is the matrix $(\delta_{1,1})$, so that trace (xay) = 1 = trace (ayx). Since the trace is N-linear there is a $z \in \Sigma$ with trace $(az) = rt(az) \neq 0$. This proves that ψ is a monomorphism. By a count of dimensions, ψ is an isomorphism.

<u>Lemma 2.6</u>: Let R be a Dedekind domain with quotient field K and let Σ be a central simple K-algebra. Then any order A over R in Σ is a finitely generated R-module.

<u>Proof</u>: Let $u_1,\ldots,u_n$ be a K-basis for Σ in A and let L be the free R-module generated by $u_1,\ldots,u_n$. Let $f_i \in Hom_K(\Sigma,K)$ be defined by $f_i(u_j) = \delta_{i,j}$. Since ψ is an isomorphism there are elements $v_1,\ldots,v_n$ in Σ with $f_i(x) = rt(v_i x)$ for all $x \in \Sigma$. Let L^c be the free R-submodule of Σ generated by $v_1,\ldots,v_n$.

Observe that $L^c = \{x \in \Sigma \mid rt(xL) \subseteq R\}$. Since $L \subseteq A$ we know that $A \cdot L \subseteq A$. But $rt(A) \subseteq R$ so $rt(AL) \subseteq R$. Therefore $A \subseteq L^c$ and since R is noetherian this implies A is finitely generated.

Lemma 2.7: Let R be a local principal ideal domain with quotient field K, and let Σ be a central simple K-algebra. If A and B are R-orders in Σ with B central separable as an R-algebra then there is an invertible element $t \in \Sigma$ with $A \subseteq t^{-1}Bt$. In particular, B is a maximal order.

Proof: Let $F = \{x \in \Sigma \mid xA \subseteq B\}$. It is clear that F is a left ideal in B since $1 \in A$. Let $x_1, \ldots, x_n$ be a generating set for A over R and let $y_1, \ldots, y_n$ be a generating set for B over R.

There are elements $\alpha_{ij} \in K$ with

$$y_i = \Sigma_{j=1}^{n} \alpha_{ij} x_j.$$

Each $\alpha_{ij} = \beta_{ij}/\gamma_{ij}$ with $\beta_{ij}, \gamma_{ij} \in R$ so letting $c = \Pi_{i,j} \gamma_{ij}$ we see that $0 \neq c \in R$ and $c \cdot A \subseteq B$. Therefore $F \cap R \neq \{0\}$. Now F is a submodule of the finitely generated projective R-module B so F is finitely generated, torsion-free and therefore projective. By Lemma 2.3 $F = B \cdot t$ for some $t \in B$. Since $F \cap R \neq 0$, the left ideal generated by t in Σ contains a unit so is all of Σ. Therefore t is a unit in Σ. Now Bt is a right A-module so $tA \subseteq Bt$ or $A \subseteq t^{-1}Bt$.

In particular, if A is a maximal order containing B then $t^{-1}Bt = A$. Inner automorphism by t on Σ is an algebra isomorphism and one can check that the image of a maximal order under a K-algebra isomorphism of Σ is a maximal order. Therefore B is a maximal order.

Lemma 2.8: Let A be a maximal order over the Dedekind domain R in the central simple algebra Σ over the quotient field K of R. If $\mathfrak{p}$ is a prime ideal in R, then $R_{\mathfrak{p}} \otimes A$ is a maximal order over $R_{\mathfrak{p}}$ in Σ.

Proof: Identify both $R_{\mathfrak{p}}$ and $R_{\mathfrak{p}} \otimes A$ with R-subalgebras of Σ by viewing $R_{\mathfrak{p}}$ as a partial ring of quotients of R in K and $R_{\mathfrak{p}} \otimes A$ as the R-subalgebra of Σ generated by $R_{\mathfrak{p}}$ and A. It is clear that $R_{\mathfrak{p}} \otimes A$ is an order over $R_{\mathfrak{p}}$ in Σ. Suppose C is an order over $R_{\mathfrak{p}}$ in Σ containing $R_{\mathfrak{p}} \otimes A$. Let $F = \{x \in \Sigma \mid x \cdot C \subseteq R_{\mathfrak{p}} \otimes A\}$ and observe that F is a two-sided ideal in $R_{\mathfrak{p}} \otimes A$. Also, $F \cap A$ is a right ideal in A and since $F \cap R \neq 0$ (see the proof of Lemma 2.7) we have $K(F \cap A)$ is a right ideal in Σ containing an element of K so $K(F \cap A) = \Sigma$. Let $D_r(F \cap A) = \{x \in \Sigma \mid (F \cap A)x \subseteq (F \cap A)$. Since $F \cap A$ is finitely generated as an R-module and $K(F \cap A) = \Sigma$ one can check that $D_r(F \cap A)$ is an R-order in Σ and $A \subseteq D_r(F \cap A)$. Therefore $A = D_r(F \cap A)$.

Now we show $C \subseteq R_{\mathfrak{p}} \otimes A$. Let $\delta \in C$, then $F\delta \subseteq R_{\mathfrak{p}} \otimes A$. Therefore $(F \cap A)\delta \subseteq R_{\mathfrak{p}} \otimes A$. Since $(F \cap A)\delta$ is a finitely generated R-module, there is an element $y \in R - \mathfrak{p}$ so that $(F \cap A)\delta y \subseteq A$. Also $(F \cap A)\delta y \subseteq F$ from the definition of F. Therefore $\delta y \in D_r(F \cap A)$ so $\delta y \in A$. Therefore $\delta \in R_{\mathfrak{p}} \otimes A$ which proves the lemma.

If R is a Dedekind domain, then by Lemma 2.2 one can identify $B(R)$ and $B(R_{\mathfrak{p}})$ with corresponding subgroups of $B(K)$ where $R_{\mathfrak{p}}$ is the localization of R at the prime ideal $\mathfrak{p}$ and K is the quotient field of R. Under this identification $B(R)$ is a subgroup of $B(R_{\mathfrak{p}})$.

<u>Theorem 2.9</u>: Let R be a Dedekind domain. Then

$$B(R) = \bigcap_{\mathfrak{p}} B(R_{\mathfrak{p}})$$

where $\mathfrak{p}$ ranges over the prime ideals of R.

<u>Proof</u>: The remarks preceding the theorem show $B(R) \subseteq \bigcap_{\mathfrak{p}} B(R_{\mathfrak{p}})$. To prove the other inclusion let Σ be a central separable algebra over the quotient field K of R. Assume that for each prime ideal $\mathfrak{p}$ of R there is a central separable $R_{\mathfrak{p}}$ algebra $B_{\mathfrak{p}}$ with $[K \otimes_{R_{\mathfrak{p}}} B_{\mathfrak{p}}] = [\Sigma]$ in $B(K)$. Let D be the division algebra equivalent to Σ in $B(K)$ whose existence is given by Corollary 1.3. Let A be a maximal order over R in D. By Lemma 2.6 A is a finitely generated R-module and since R is a Dedekind domain A is projective over R. Moreover it is easy to check that $K \otimes_R A \simeq D$ so for each prime ideal $\mathfrak{p}$ of R there is a finitely generated free $R_{\mathfrak{p}}$-module E with

$$K \otimes_{R_{\mathfrak{p}}} (R_{\mathfrak{p}} \otimes A) \otimes_{R_{\mathfrak{p}}} \mathrm{Hom}_{R_{\mathfrak{p}}}(E,E) \simeq K \otimes_{R_{\mathfrak{p}}} B_{\mathfrak{p}}.$$

To check this isomorphism observe that both sides are just $n \times n$ matrices over D where E is chosen with $\mathrm{Rank}_{R_{\mathfrak{p}}}(E) = n$.

By Lemma 2.8, $R_{\mathfrak{p}} \otimes A$ is a maximal order over $R_{\mathfrak{p}}$ in D. Let $A' = (R_{\mathfrak{p}} \otimes A) \otimes_{R_{\mathfrak{p}}} \mathrm{Hom}_{R_{\mathfrak{p}}}(E,E)$ and view the elements of A' as $n \times n$ matrices with entries in $R_{\mathfrak{p}} \otimes A$. We claim A' is a maximal order over $R_{\mathfrak{p}}$ in $K \otimes_{R_{\mathfrak{p}}} B_{\mathfrak{p}}$. Let C be an order over $R_{\mathfrak{p}}$ containing A' and let $[d_{ij}]$ be an $n \times n$ matrix over D belonging to C. Since the elementary matrices are in A', by pre and post multiplying $[d_{ij}]$ by appropriate elements of A' we can find for any k, j another element of C whose diagonal entries are the element d_{ij} and whose off-diagonal entries are 0. Since $R_{\mathfrak{p}} \otimes A$ is a maximal order, $C \cap R_{\mathfrak{p}} \otimes A = R_{\mathfrak{p}} \otimes A$ so $d_{ij} \in R_{\mathfrak{p}} \otimes A$ for all i, j and $A' \simeq C$. Therefore A' is a maximal order over $R_{\mathfrak{p}}$ in $K \otimes_{R_{\mathfrak{p}}} B_{\mathfrak{p}}$. By Lemma 2.7 $A' \simeq B_{\mathfrak{p}}$ under an inner automorphism of Σ so A' is a central separable $R_{\mathfrak{p}}$-algebra. Corollary 2.1.9 thus implies $R_{\mathfrak{p}} \otimes A$ is central separable. Since $R_{\mathfrak{p}} \otimes A$ is a central separable $R_{\mathfrak{p}}$-algebra for every prime ideal $\mathfrak{p}$ of R by Theorem 2.7.1 we know A is a central separable R-algebra. Moreover $[K \otimes A] = [\Sigma]$ in B(K) so $B(R) = \bigcap_{\mathfrak{p}} B(R_{\mathfrak{p}})$.

§ 3. Remarks

At this point we pass to a series of remarks, conjectures, and references to the literature on the subject. We make no claim for completeness but will try to bring out those problems which have interested us and which (we hope) will interest others.

The basic facts concerning the Brauer group of a commutative ring were given by M. Auslander and O. Goldman in [7]. A fundamental problem is the computation of the Brauer group functor. Given a commutative ring R, what is B(R)? Given a ring homomorphism h from R to another commutative ring S, what is B(h)? A related problem is to give the algebraic structure of the elements of B(R). What is the structure of a central separable R-algebra A? In order to give more complete answers to these questions the Galois theory and Galois cohomology of Chapter III and Chapter IV must be extended.

The Galois theory presented in Chapter III can be generalized in several directions. The general Galois theory of rings has been a popular area for research in the past twenty years. The fundamental theorem of Galois theory (Theorem 3.1.1) has been generalized in a series of papers by O. Villamayor and D. Zelinsky [92], [93], to commutative rings with any number of idempotents. At the same time, a series of authors (T. Kanzaki [60], F. DeMeyer [24], [25], L. Childs [18], H. Kreimer [68], [69], Y. Miyashita [74], [75], T. Nakayama [78]) have successively carried out generalizations of the fundamental theorem in non-commutative contexts. Attempts to generalize the Galois theory in the inseparable case have been made by S. Chase and M. Sweedler [17]. In their work a Hopf algebra replaces the usual Galois group. The imbedding theorem (Theorem 3.2.8) has been generalized by O. Villamayor [91], who has eliminated the condition on idempotents. There has, however, been no successful generalization of the separable closure and infinite Galois theory as the presence of proper idempotents appears to cause more fundamental difficulties here. (F. DeMeyer [31]) There has also been no successful attempt to eliminate the condition on idempotents from the theorems in Section 4 on separable polynomials. Classification schemes for all Galois extensions of a given commutative ring with abelian Galois group has been presented by D. K. Harrison [16] and generalized by M. Orzeck [80].

The Galois cohomology of Chapter IV was originally presented in a much more general form by S. Chase and A. Rosenberg [16], and extended by L. N. Childs [20]. It was given essentially as we have in unpublished work of M. Auslander and A. Brumer, and later was published by T. Kanzaki [65].

Aside from the results we have already presented, what are some answers to the three questions we raised at the beginning of this section? F. DeMeyer [26] showed that if R is a local ring and Ω is the separable closure of R then $B(\Omega) = 0$. A corollary of this result is that every central separable R-algebra has a splitting ring which is a Galois extension of R so the Galois cohomology (Theorem 4.1.1) implies B(R) is a torsion group. If R is a Dedekind domain with quotient field K then by Lemma 2.2 B(R) is a subgroup of B(K), thus B(R) is a torsion group.

Additional information on the group theoretic structure of B(R) can be found in Section I of Grothendieck [43].

S. Endo and T. Watanabe [39] have shown that if R is a noetherian ring any central separable R-algebra A can be split by an extension S of R but in their construction S is usually neither finite nor projective over R. G. Szeto [88] has shown that if G is a finite group of order n, if RG is a separable group algebra and if $S = R[x]/(x^n-1)$ then SG is a direct sum of central separable algebras in the zero class of B(S).

Problem 1. Which commutative rings R have the property that any central separable R-algebra A has a finitely generated projective splitting ring?

An analogous problem for class groups is the following:

Problem 2. Which commutative rings R have the following property: if $|E| \in P(R)$ there is a finite projective extension S of R with $|S \otimes E| = |S| \in P(S)$. There are several papers in the literature which discuss these two problems.

In order to compute the Brauer group one often uses the functorial properties of B( ) to reduce the problem. For example M. Auslander and O. Goldman [7] prove that if k is a perfect field then $B(k) \simeq B(k[x])$ under the map induced from the natural imbedding of k in k[x]. A. Madgid [73] has shown that if R is a Boolean ring then $B(R) = 0$. He does this by employing a construction similar to the prime ideal spectrum constructed in Section 4 of Chapter 1 except applied to the idempotent generated ideals of R. Given commutative rings R and S together with a ring homomorphism h from R to S one wishes to compute B(h). We showed (Lemma 2.1) that if R is a Dedekind domain with quotient field S and h is the natural imbedding of R in S then B(h) is a monomorphism.

Problem 3. For which integral domains R with quotient field K is the induced homomorphism from B(R) to B(K) a monomorphism? This problem was given by M. Auslander and O. Goldman in [7].

Now let R be a local ring with maximal ideal $\mathfrak{m}$. If R is complete in its $\mathfrak{m}$-adic topology the natural homomorphism from R to $R/\mathfrak{m}$ induces an isomorphism from B(R) to $B(R/\mathfrak{m})$. (Auslander and Goldman [7], Azumaya [8]) For general local rings the homomorphism from B(R) to $B(R/\mathfrak{m})$ need not be either one-to-one nor onto. There are some special sorts of local rings which arise in algebraic geometry for which a computation of this homomorphism would be interesting (see A. Grothendieck [43], [44]).

Problem 4. Given a local ring R with maximal ideal $\mathfrak{m}$, compute the kernel of the induced homomorphism from B(R) to $B(R/\mathfrak{m})$.

The results we have presented in Chapter V are not in their most general form. For an up-to-date treatment of the theory of orders we refer the reader to R. M. Fossum [41].

Little is known about the structure of separable algebras in the general case. We might ask the following series of questions.

Problem 5. For which commutative rings R are the following true?

(a) If A is a central separable R-algebra then there is a unique central separable R-algebra D equivalent to A in B(R) so that D has no idempotents other than 0 and 1 and $A^o = \mathrm{Hom}_D(M,M)$ for some D-progenerator M.

(b) If A is a central separable R-algebra then every indecomposable projective left A-module is of the form $M \otimes X$ for M a fixed indecomposable projective A-module and $|X| \in P(R)$.

In Theorem 1.1 we have shown that (a) and (b) are equivalent. H. Bass has pointed out the existence of principal ideal domains where (a) and (b) fail (Bass [10]). F. DeMeyer [29] has shown that if R is a local ring or the ring of polynomials in one variable over a perfect field then (a) and (b) both hold. L. N. Childs has presented several interesting examples in [19].

Finally, let us discuss one avenue by which the theory of separable algebras can be extended to more general algebras. If A is a finite dimensional algebra with radical N over a field R with A/N separable over R then the Wedderburn-Malcev Theorem asserts that there is a separable R-subalgebra S of A with $S + N = A$ (vector space direct sum) and S unique up to inner automorphism by a unit in A of the form $1 + n$, $n \in N$. Generalizations of this result would give information about finitely generated algebras over a commutative ring. G. Azumaya [8], W. C. Brown [11], and E. C. Ingraham [56], [57] have already given some generalizations of parts of the Wedderburn-Malcev theorem to finitely generated algebras over special kinds of local rings, Dedekind domains, and polynomial rings.

Problem 6. Find commutative rings R to which the Wedderburn-Malcev theorem can be extended.

Exercises

1. Let R be a local ring and let A be a central separable R-algebra. Then

1. Any two indecomposable finitely generated projective A-modules are isomorphic.
2. There is a unique (up to isomorphism) central separable R-algebra D with no idempotents other than 0 and 1 equivalent

to A in B(R).

3. A is the ring of all n × n matrices with entries from D.

Hint: Let $\mathfrak{p}$ be the maximal ideal of R, let M,N be finitely generated indecomposable projective A-modules with $\mathrm{Rank}_R(M) \geq \mathrm{Rank}_R(N)$. Then there is an A-homomorphism $\overline{f}$ from $M/\mathfrak{p}M$ onto $N/\mathfrak{p}N$. Lift $\overline{f}$ to an isomorphism f from M onto N.

2. (G. Sezto) Let R be a commutative ring, let G be a finite group, and let RG be a separable group algebra (see Chapter 2, Section 1). Let M be an RG-module which is finitely generated projective over R (equivalently RG) and let $f_1,\ldots,f_n$, $x_1,\ldots,x_n$ be a dual basis for M over R. Define $t_M: G \to R$ by

$$t_M(g) = \sum_{i=1}^{n} f_i(gx_i)$$

We call t_M the character on G afforded by M.

Prove that

1. t_M is independent of the dual basis chosen for M.
2. t_M is a class function on G (ie., $t_M(ghg^{-1}) = t_M(h)$ for all g, h $\in$ G.
3. If M = RG compute t_M explicitly.
4. Let n be the number of elements in G, and assume the only idempotents in R are 0 and 1.

Prove that $S = R[x]/(x^n-1)$ is separable over R.

5. Generalize the Brauer splitting theorem by showing $S \otimes RG$ is a finite direct sum of elements in the zero class of B(S).

Hint: When the characteristic of R is 0 observe that $Z(1/n, \sqrt[n]{1}) \subseteq S$ and use Lemma 4.2.2 together with the classical Brauer theorem. Handle the finite characteristic case in a similar way.

3. Let D be the algebra of real quaternions. D is a division algebra over the real numbers with basis 1,i,j,k where

$$i^2 = j^2 = k^2 = -1 \text{ and } ij = -ji = k$$

Let $R = Z(\sqrt{2})$ (Z = ring of integers) and let

$$\alpha = 1+i/2,\ \beta = i+j/2,\ \alpha\beta = (1+i+j+k)/2\ ,$$

Let $A = R\cdot 1 + R\alpha + R\beta + R\alpha\beta$. Prove that R is a principal ideal domain. Prove that A is a central separable R-algebra, and A is not in the zero class of $B(R)$. From algebraic number theory, R is separably closed so conclude A cannot be split by a separable projective extension of R.

4. Let R be a Dedekind domain with quotient field K and let A be an R-order in the central simple K-algebra Σ. Prove A is separable over R if and only if rt generated $\mathrm{Hom}_R(A,R)$ freely as a right A-module.

BOOKS

A.) Artin, E. Galois Theory. Notre Dame Mathematical Lectures No. 2 (1942).

B.) Artin, E., Nesbitt, C.J., Thrall, R.M. Rings with Minimum Condition. University of Michigan Press (1944).

C.) Bass, H. Algebraic K-Theory. Benjamin (1968).

D.) Bourbaki, N. Eléments de Mathematique, Livre II, Chapitre 8, Modules et Anneaux Semi-simples. Hermann (1958).

E.) Bourbaki, N. Eléments de Mathematique, Algèbre Commutative, Chapters 1 and 2. Hermann (1961).

F.) Cartan, H. and Eilenberg, S. Homological Algebra. Princeton Mathematical Series No. 19 (1956).

G.) Curtis, C. W. and Reiner, I. Representation Theory of Finite Groups and Associative Algebras. Interscience (1962).

H.) Lang, S. Algebra. Addison-Wesley (1965).

I.) Northcott, D. G. An Introduction to Homological Algebra. Cambridge (1962).

J.) Van der Waerden, B. L. Modern Algebra, Vol. II. Ungar (1950).

K.) Weiss, E. Algebraic Number Theory. McGraw-Hill (1963).

L.) Zariski, O. and Samuel, P. Commutative Algebra, Vol. I. Van Nostrand (1958).

BIBLIOGRAPHY

1. Anderson, F.W., Endomorphism rings of projective modules, Math. Z. 111 (1969), 322-332.

2. Artin, M., On Azumaya algebras and finite dimensional representations of rings, J. Algebra 11 (1969), 532-563. MR 39#4217.

3. Auslander, B., The Brauer group of a ringed space, J. Algebra 4 (1966), 220-272. MR 33#7262.

4. Auslander, M. and Buchsbaum, D.A., On ramification theory in noetherian rings, Amer. J. Math. 81 (1959), 749-765. MR 21#5659.

5. Auslander, M. and Brumer, A., Brauer groups of discrete valuation rings, Nederl. Akad. Wetensch. Proc. Ser. A71 = Indag. Math. 30 (1968), 286-296. MR 37# 4051.

6. Auslander, M. and Goldman, O., Maximal orders, Trans. Amer. Math. Soc. 97(1960), 1-24. MR 22#8034.

7. Auslander, M. and Goldman, O., The Brauer group of a commutative ring, Trans. Amer. Math. Soc. 97 (1960), 367-409. MR 22#12130.

8. Azumaya, G., On maximally central algebras, Nagoya Math. J. 2 (1951), 119-150. MR 12 p. 669.

9. Barr, M. and Knus, M.-A., Extensions of derivations, Forschungsinstitut für mathematik, E.T.H., Zürich (1970).

10. Bass, H., K-theory and stable algebra, Publ. I.H.E.S. No. 22 (1964), 5-60.

11. Brown, W.C., Strong inertial coefficient rings, Mich. Math. J. 17 (1970), 73-84.

12. Brown, W.C., Some splitting theorems for algebras over commutative rings, (to appear).

13. Brown, W.C., and Ingraham, E.C., A characterization of semi-local inertial coefficient rings, Proc. Amer. Math. Soc., Sept. 1970.

14. Brumer, A., and Rosen, M., On the size of the Brauer group, Proc. Amer. Math. Soc. 19 (1968), 707-711. MR 37#1362.

15. Chase, S.U., Galois objects and extensions of Hopf algebras

16. Chase, S.U., Harrison, D.K. and Rosenberg, A., Galois theory and cohomology of commutative rings, Mem. Amer. Math. Soc. 52 (1965), MR 33#4117-4119.

17. Chase, S.U. and Sweedler, M.E., Hopf algebras and Galois theory, Mem. Amer. Math. Soc.

18. Childs, L.N., A note on the fixed ring of a Galois extension, Osaka J. Math. 4 (1967), 173-176. MR 36#1490.

19. Childs, L.N., On projective modules and automorphisms of central separable algebras, Canad. J. Math. 21 (1969), 44-53. MR 38#5837.

20. Childs, L.N., Group extensions and cohomology and Galois objects with normal basis (to appear).

21. Childs, L.N., The exact sequence of low degree and normal algebras, Bull. Amer. Math. Soc. 76 (1970), 1121-1124.

22. Childs, L.N. and DeMeyer, F.R., On automorphisms of separable algebras, Pacific J. Math. 23 (1967), 25-34. MR 36# 213.

23. Cunningham, R.S., Strongly separable pairings of rings, Trans. Amer. Math. Soc. 148 (1970), 399-416.

24. DeMeyer, F.R., Some notes on the general Galois theory of rings, Osaka J. Math. 2 (1965), 117-127. MR 32#128.

25. DeMeyer, F.R., Galois theory in separable algebras over commutative rings, Illinois J. Math. 10 (1966), 287-295. MR 33#149.

26. DeMeyer, F.R., The Brauer group of some separably closed rings, Osaka J. Math. 3 (1966), 201-204. MR 35#2881.

27. DeMeyer, F.R., The trace map and separable algebras, Osaka J. Math. 3 (1966), 7-11. MR 37#4122.

28. DeMeyer, F.R., Another proof of the fundamental theorem of Galois theory, Amer. Math. Monthly 75 (1968), 720-724. MR 39#5525.

29. DeMeyer, F.R., Projective modules over central separable algebras, Canad. J. Math. 21 (1969), 39-43. MR 38#3299.

30. DeMeyer, F.R., On automorphisms of separable algebras II, Pacific J. Math. 32 (1970), 621-631.

31. DeMeyer, F.R., Separable polynomials over a commutative ring. (To appear).

32. Deuring, M., Zur theorie der Idealklassen in algebraischen Funktionkörpern, Math. Ann. 106 (1932), 103-106.

33. Eilenberg, S. and Nakayama, T., On the dimension of modules and algebras II, Nagoya Math. J. 9 (1955), 1-16. MR 17 p.453.

34. Elkins, B.L., Characterization of separable ideals, Pacific J. Math. (To appear).

35. Endo, S., Completely faithful modules and quasi Frobenius algebras, J. Math. Soc. Japan 19 (1967), 437-456. MR 37#1406.

36. Endo, S., On ramification theory in projective orders, Nagoya Math. J. 36 (1969), 121-141.

37. Endo, S. and Watanabe, Y., The centers of semi-simple algebras over a commutative ring, Nagoya Math. J. 30 (1967), 285-293. MR#4256.

39. Endo, S. and Watanabe, Y., On separable algebras over a commutative ring, Osaka J. Math. 4 (1967), 233-242. MR 37#2796.

40. Fossum, R., The Noetherian different of projective orders, J. Reine Angew. Math. 224 (1966), 207-218. MR 36#5119.

41. Fossum, R., Maximal orders over Krull domains, J. Algebra 10 (1968), 321-332. MR 38#2130.

42. Garfunkel, G. and Orzeck, M., Galois extensions as modules over the group ring. (To appear).

43. Grothendieck, A., Le groupe de Brauer I: Algèbres d'Azumaya et interprétations diverses, Séminaire Bourbaki 1964/65 No. 290.

44. Grothendieck, A., Le groupe de Brauer II: Théorie cohomologique, Séminaire Bourbaki 1965/66 No. 297.

45. Harada, M., Some criteria for hereditarity of crossed products, Osaka J. Math. 1 (1964), 69-80. MR 30#4785.

46. Harada, M., Supplementary results on Galois extension, Osaka J. Math. 2 (1965), 343-350. MR 33#151.

47. Harada, M., Correction to: Supplementary results on Galois extension, Osaka J. Math. 3 (1966), 187. MR 33#4103.

48. Harada, M. and Kanzaki, T., Galois theory of rings (Japanese) Sûgaku 18 (1966), 144-159. MR 36#6459.

49. Harada, M., Note on Galois extension over the center, Revista Un. Mat. Argentina 24 (1968), 91-96.

50. Hattori, A., Semisimple algebras over a commutative ring, J. Math. Soc. Japan 15 (1963), 404-419. MR 28#2125.

51. Hattori, A., On strongly separable algebras, Osaka J. Math. 2 (1965), 369-372. MR 34#203.

52. Hirata, K. and Sugano, K., On semisimple extensions and separable extensions over non-commutative rings, J. Math. Soc. Japan 18 (1966), 360-373. MR 34#208.

53. Hirata, K., Some types of separable extensions of rings, Nagoya Math. J. 33 (1968), 107-115. MR 38#4524.

54. Hirata, K., Separable extensions and centralizers of rings, Nagoya Math. J. 35 (1969), 31-45. MR 39#5636.

55. Hongan, M. and Nagahara, T., A note on separable extensions of commutative rings, Math. J. Okayama U. 14 (1969), 13-15.

56. Ingraham, E.C., Inertial subalgebras of algebras over commutative rings, Trans. Amer. Math. Soc. 124 (1966), 77-93. MR 34#209.

57. Ingraham, E.C., Inertial subalgebras of complete algebras, J. Algebra 15 (1970), 1-11.

58. Janusz, G.J., Separable algebras over commutative rings, Trans. Amer. Math. Soc. 122 (1966),461-479. MR 35#1585.

59. Kanzaki, T., A type of separable algebras, J. Math. Osaka City Univ. 13 (1962), 39-43. MR 27#4838.

60. Kanzaki, T., On commutor rings and Galois theory of separable algebras, Osaka J. Math. 1 (1964), 103-115; correction ibid 1 (1964), 253. MR 29#5865.

61. Kanzaki, T., Special type of separable algebra over a commutative ring, Proc. Japan Acad. 40 (1964), 781-786. MR 31#3455.

62. Kanzaki, T., On Galois algebra over a commutative ring, Osaka J. Math. 2 (1965),309-317. MR 33#150.

63. Kanzaki, T., A note on abelian Galois algebra over a commutative ring, Osaka J. Math. 3 (1966), 1-6. MR 33#5680.

64. Kanzaki, T., On Galois extensions of rings, Nagoya Math. J. 27 (1966), 43-49. MR 35#2928.

65. Kanzaki, T., On generalized crossed product and Brauer group, Osaka J. Math. 5 (1968), 175-188. MR 39#2809.

66. Knus, M-A., Sur le théoreme de Skolem-Noether et sur les dérivations des algèbres d'Azumaya, C.R. Acad. Sc. Paris 270 (1970), 637-639.

67. Knus, M-A., Un théoreme de cancellation pour les algèbres d'Azumaya, Comm. Helv. Math. (To appear).

68. Kreimer, H.F., A Galois theory for non-commutative rings, Trans. Amer. Math. Soc. 127 (1967), 29-41. MR 35#6718.

69. Kreimer, H.F., Galois theory for non-commutative rings and normal bases, Trans. Amer. Math. Soc. (1967), 42-49. MR 35#6719.

70. Kreimer, H.F., A note on the outer Galois theory of rings, Pacific J. Math. 31 (1969), 417-432.

71. Kreimer, H.F., Outer Galois theory for separable algebras, Pacific J. Math. 32 (1970), 147-156.

72. Magid, A., Commutative algebras of Hochschild dimension one, Proc. Amer. Math. Soc. 24 (1970), 530-532.

73. Magid, A., Pierce's representation and separable algebras, Illinois J. Math. (To appear).

74. Miyashita, Y., Locally finite outer Galois theory, J. Fac. Sci. Hokkaido Univ. Ser. I. 20 (1967), 1-26. MR 38#1126.

75. Miyashita, Y., Finite outer Galois theory of non-commutative rings, J. Fac. Sci. Hokkaido Univ. Ser. I. 19 (1966), 114-134. MR 35#1638.

76. Miyashita, Y., Galois extensions and crossed products, J. Fac. Sci. Hokkaido Univ. Ser. I 20 (1968), 122-134. MR 39#262.

77. Nagahara, T., A note on Galois theory of commutative rings, Proc. Amer. Math. Soc. 18 (1967), 334-340. MR 34#7580.

78. Nakayama, T., On a generalized notion of Galois extensions of a ring, Osaka Math. J. 15 (1963), 11-23. MR 27#1478.

79. Onodera, T., A characterization of strongly separable algebras, J. Fac. Sci. Hokkaido Univ. Ser. I 19 (1966), 71-73. MR 33#5678.

80. Orzech, M., A cohomological description of abelian Galois extensions, Trans. Amer. Math. Soc. 137 (1969), 481-499.

81. Roggenkamp, K.W., A remark on separable orders, Canad. Math. Bull. 12 (1969), 453-56. MR 40#175.

82. Rosenberg, A. and Zelinsky, D., On Amitsur's complex, Trans. Amer. Math. Soc. 97 (1960), 327-356. MR 22#12129.

83. Rosenberg, A. and Zelinsky, D., Automorphisms of separable algebras, Pac. J. Math. 11 (1961), 1109-1117. MR 26#6215.

84. Roy, A. and Sridharan, R., Derivations in Azumaya algebras, J. Math. Kyoto Univ. 7 (1967), 161-167. MR 36#5125.

85. Roy, A. and Sridharan, R., Higher derivations and central simple algebras, Nayoya Math. J. 32 (1968), 21-30. MR 37#5255.

86. Sugano, K., Note on semisimple extensions and separable extensions, Osaka J. Math. 4 (1967), 265-270. MR 37#1412.

87. Strooker, J.R., Faithfully projective modules and clean algebras, Dissertation, U. of Utrecht, Utrecht 1965, J.J. Groen and Zoon, N.V. Leiden 1965. MR 36#206.

88. Szeto, G. On the Brauer Splitting Theorem, Pacific J. Math. 31 (1969), 505-512.

89. Takeuchi, Y., On Galois extensions over commutative rings, Osaka J. Math. 2 (1965), 137-145. MR 32#129.

90. Takeuchi, Y., Infinite outer Galois theory of non-commutative rings, Osaka J. Math. 3 (1966), 195-200. MR 35#1639.

91. Villamayor, O.E., Separable algebras and Galois extensions, Osaka J. Math. 4 (1967), 161-171. MR 37#4064.

92. Villamayor, O.E. and Zelinsky, D., Galois theory for rings with finitely many idempotents, Nagoya Math. J. 27 (1966) 721-731. MR 34#5880.

93. Villamayor, O.E. and Zelinsky, D., Galois theory with infinitely many idempotents, Nagoya Math. J. 35 (1969), 83-98. MR 39#5555.

94. Watanabe, Y., The Dedekind different and the homological different, Osaka J. Math. 4 (1967), 227-231. MR 37#2795.

95. Yuan, S., On the Brauer group of local fields, Ann. of Math. (2) 82 (1965), 434-444. MR 32#2449.

INDEX

cture Notes in Mathematics

r erschienen/Already published

: J. Wermer, Seminar über Funktionen-Algebren. IV, 30 Seiten. DM 3,80 / $ 1.10

2: A. Borel, Cohomologie des espaces localement compacts s. J. Leray. IV, 93 pages. 1964. DM 9,– / $ 2.60

: J. F. Adams, Stable Homotopy Theory. Third edition. IV, 78 pages. DM 8,– / $ 2.20

: M. Arkowitz and C. R. Curjel, Groups of Homotopy Classes. evised edition. IV, 36 pages. 1967. DM 4,80 / $ 1.40

5: J.-P. Serre, Cohomologie Galoisienne. Troisième édition. 14 pages. 1965. DM 18,– / $ 5.00

: H. Hermes, Term Logic with Choise Operator. III, 55 pages. DM 6,– / $ 1.70

: Ph. Tondeur, Introduction to Lie Groups and Transformation s. Second edition. VIII, 176 pages. 1969. DM 14,– / $ 3.80

: G. Fichera, Linear Elliptic Differential Systems and Eigenvalue ems. IV, 176 pages. 1965. DM 13,50 / $ 3.80

. P. L. Ivănescu, Pseudo-Boolean Programming and Applications. pages. 1965. DM 4,80 / $ 1.40

0: H. Lüneburg, Die Suzukigruppen und ihre Geometrien. VI, ten. 1965. DM 8,– / $ 2.20

: J.-P. Serre, Algèbre Locale. Multiplicités. Rédigé par P. Gabriel. de édition. VIII, 192 pages. 1965. DM 12,– / $ 3.30

: A. Dold, Halbexakte Homotopiefunktoren. II, 157 Seiten. 1966. ,– / $ 3.30

3: E. Thomas, Seminar on Fiber Spaces. IV, 45 pages. 1966. 80 / $ 1.40

: H. Werner, Vorlesung über Approximationstheorie. IV, 184 Seid 12 Seiten Anhang. 1966. DM 14,– / $ 3.90

5: F. Oort, Commutative Group Schemes. VI, 133 pages. 1966. 80 / $ 2.70

6: J. Pfanzagl and W. Pierlo, Compact Systems of Sets. IV, jes. 1966. DM 5,80 / $ 1.60

: C. Müller, Spherical Harmonics. IV, 46 pages. 1966. – / $ 1.40

: H.-B. Brinkmann und D. Puppe, Kategorien und Funktoren. 7 Seiten, 1966. DM 8,– / $ 2.20

: G. Stolzenberg, Volumes, Limits and Extensions of Analytic es. IV, 45 pages. 1966. DM 5,40 / $ 1.50

): R. Hartshorne, Residues and Duality. VIII, 423 pages. 1966. ,– / $ 5.50

: Seminar on Complex Multiplication. By A. Borel, S. Chowla, erz, K. Iwasawa, J.-P. Serre. IV, 102 pages. 1966. DM 8,– /$ 2.20

2: H. Bauer, Harmonische Räume und ihre Potentialtheorie. Seiten. 1966. DM 14,– / $ 3.90

: P. L. Ivănescu and S. Rudeanu, Pseudo-Boolean Methods for it Programming. 120 pages. 1966. DM 10,– / $ 2.80

4: J. Lambek, Completions of Categories. IV, 69 pages. 1966. 30 / $ 1.90

: R. Narasimhan, Introduction to the Theory of Analytic Spaces. pages. 1966. DM 10,– / $ 2.80

6: P.-A. Meyer, Processus de Markov. IV, 190 pages. 1967. – / $ 4.20

7: H. P. Künzi und S. T. Tan, Lineare Optimierung großer e. VI, 121 Seiten. 1966. DM 12,– / $ 3.30

: P. E. Conner and E. E. Floyd, The Relation of Cobordism to ries. VIII, 112 pages. 1966. DM 9,80 / $ 2.70

9: K. Chandrasekharan, Einführung in die Analytische Zahlen-. VI, 199 Seiten. 1966. DM 16,80 / $ 4.70

: A. Frölicher and W. Bucher, Calculus in Vector Spaces without X, 146 pages. 1966. DM 12,– / $ 3.30

: Symposium on Probability Methods in Analysis. Chairman. appos. IV, 329 pages. 1967. DM 20,– / $ 5.50

: M. André, Méthode Simpliciale en Algèbre Homologique et e Commutative. IV, 122 pages. 1967. DM 12,– / $ 3.30

3: G. I. Targonski, Seminar on Functional Operators and ons. IV, 110 pages. 1967. DM 10,– / $ 2.80

: G. E. Bredon, Equivariant Cohomology Theories. VI, 64 pages. M 6,80 / $ 1.90

: N. P. Bhatia and G. P. Szegö, Dynamical Systems. Stability and Applications. VI, 416 pages. 1967. DM 24,– / $ 6.60

: A. Borel, Topics in the Homology Theory of Fibre Bundles. pages. 1967. DM 9,– / $ 2.50

Vol. 37: R. B. Jensen, Modelle der Mengenlehre. X, 176 Seiten. 1967. DM 14,– / $ 3.90

Vol. 38: R. Berger, R. Kiehl, E. Kunz und H.-J. Nastold, Differentialrechnung in der analytischen Geometrie IV, 134 Seiten. 1967 DM 12,– / $ 3.30

Vol. 39: Séminaire de Probabilités I. II, 189 pages. 1967. DM 14,– / $ 3.90

Vol. 40: J. Tits, Tabellen zu den einfachen Lie Gruppen und ihren Darstellungen. VI, 53 Seiten. 1967. DM 6.80 / $ 1.90

Vol. 41: A. Grothendieck, Local Cohomology. VI, 106 pages. 1967. DM 10,– / $ 2.80

Vol. 42: J. F. Berglund and K. H. Hofmann, Compact Semitopological Semigroups and Weakly Almost Periodic Functions. VI, 160 pages. 1967. DM 12,– / $ 3.30

Vol. 43: D. G. Quillen, Homotopical Algebra. VI, 157 pages. 1967. DM 14,– / $ 3.90

Vol. 44: K. Urbanik, Lectures on Prediction Theory. IV, 50 pages. 1967. DM 5,80 / $ 1.60

Vol. 45: A. Wilansky, Topics in Functional Analysis. VI, 102 pages. 1967. DM 9,60 / $ 2.70

Vol. 46: P. E. Conner, Seminar on Periodic Maps. IV, 116 pages. 1967. DM 10,60 / $ 3.00

Vol. 47: Reports of the Midwest Category Seminar I. IV, 181 pages. 1967. DM 14,80 / $ 4.10

Vol. 48: G. de Rham, S. Maumary et M. A. Kervaire, Torsion et Type Simple d'Homotopie. IV, 101 pages. 1967. DM 9,60 / $ 2.70

Vol. 49: C. Faith, Lectures on Injective Modules and Quotient Rings. XVI, 140 pages. 1967. DM 12,80 / $ 3.60

Vol. 50: L. Zalcman, Analytic Capacity and Rational Approximation. VI, 155 pages. 1968. DM 13.20 / $ 3.70

Vol. 51: Séminaire de Probabilités II. IV, 199 pages. 1968. DM 14,– / $ 3.90

Vol. 52: D. J. Simms, Lie Groups and Quantum Mechanics. IV, 90 pages. 1968. DM 8,– / $ 2.20

Vol. 53: J. Cerf, Sur les difféomorphismes de la sphère de dimension trois (Γ_4 = O). XII, 133 pages. 1968. DM 12,– / $ 3.30

Vol. 54: G. Shimura, Automorphic Functions and Number Theory. VI, 69 pages. 1968. DM 8,– / $ 2.20

Vol. 55: D. Gromoll, W. Klingenberg und W. Meyer, Riemannsche Geometrie im Großen. VI, 287 Seiten. 1968. DM 20,– / $ 5.50

Vol. 56: K. Floret und J. Wloka, Einführung in die Theorie der lokalkonvexen Räume. VIII, 194 Seiten. 1968. DM 16,– / $ 4.40

Vol. 57: F. Hirzebruch und K. H. Mayer, O (n)-Mannigfaltigkeiten, exotische Sphären und Singularitäten. IV, 132 Seiten. 1968. DM 10,80 / $ 3.00

Vol. 58: Kuramochi Boundaries of Riemann Surfaces. IV, 102 pages. 1968. DM 9,60 / $ 2.70

Vol. 59: K. Jänich, Differenzierbare G-Mannigfaltigkeiten. VI, 89 Seiten. 1968. DM 8,– / $ 2.20

Vol. 60: Seminar on Differential Equations and Dynamical Systems. Edited by G. S. Jones. VI, 106 pages. 1968. DM 9,60 / $ 2.70

Vol. 61: Reports of the Midwest Category Seminar II. IV, 91 pages. 1968. DM 9,60 / $ 2.70

Vol. 62: Harish-Chandra, Automorphic Forms on Semisimple Lie Groups X, 138 pages. 1968. DM 14,– / $ 3.90

Vol. 63: F. Albrecht, Topics in Control Theory. IV, 65 pages. 1968. DM 6,80 / $ 1.90

Vol. 64: H. Berens, Interpolationsmethoden zur Behandlung von Approximationsprozessen auf Banachräumen. VI, 90 Seiten. 1968. DM 8,– / $ 2.20

Vol. 65: D. Kölzow, Differentiation von Maßen. XII, 102 Seiten. 1968. DM 8,– / $ 2.20

Vol. 66: D. Ferus, Totale Absolutkrümmung in Differentialgeometrie und -topologie. VI, 85 Seiten. 1968. DM 8,– / $ 2.20

Vol. 67: F. Kamber and P. Tondeur, Flat Manifolds. IV, 53 pages. 1968. DM 5,80 / $ 1.60

Vol. 68: N. Boboc et P. Mustată, Espaces harmoniques associés aux opérateurs différentiels linéaires du second ordre de type elliptique. VI, 95 pages. 1968. DM 8,60 / $ 2.40

Vol. 69: Seminar über Potentialtheorie. Herausgegeben von H. Bauer. VI, 180 Seiten. 1968. DM 14,80 / $ 4.10

Vol. 70: Proceedings of the Summer School in Logic. Edited by M. H. Löb. IV, 331 pages. 1968. DM 20,– / $ 5.50

Vol. 71: Séminaire Pierre Lelong (Analyse), Année 1967 – 1968. VI, 190 pages. 1968. DM 14,– / $ 3.90

Bitte wenden / Continued

Vol. 72: The Syntax and Semantics of Infinitary Languages. Edited by J. Barwise. IV, 268 pages. 1968. DM 18,– / $ 5.00

Vol. 73: P. E. Conner, Lectures on the Action of a Finite Group. IV, 123 pages. 1968. DM 10,– / $ 2.80

Vol. 74: A. Fröhlich, Formal Groups. IV, 140 pages. 1968. DM 12,– / $ 3.30

Vol. 75: G. Lumer, Algèbres de fonctions et espaces de Hardy. VI, 80 pages. 1968. DM 8,– / $ 2.20

Vol. 76: R. G. Swan, Algebraic K-Theory. IV, 262 pages. 1968. DM 18,– / $ 5.00

Vol. 77: P.-A. Meyer, Processus de Markov: la frontière de Martin. IV, 123 pages. 1968. DM 10,– / $ 2.80

Vol. 78: H. Herrlich, Topologische Reflexionen und Coreflexionen. XVI, 166 Seiten. 1968. DM 12,– / $ 3.30

Vol. 79: A. Grothendieck, Catégories Cofibrées Additives et Complexe Cotangent Relatif. IV, 167 pages. 1968. DM 12,– / $ 3.30

Vol. 80: Seminar on Triples and Categorical Homology Theory. Edited by B. Eckmann. IV, 398 pages. 1969. DM 20,– / $ 5.50

Vol. 81: J.-P. Eckmann et M. Guenin, Méthodes Algébriques en Mécanique Statistique. VI, 131 pages. 1969. DM 12,– / $ 3.30

Vol. 82: J. Wloka, Grundräume und verallgemeinerte Funktionen. VIII, 131 Seiten. 1969. DM 12,– / $ 3.30

Vol. 83: O. Zariski, An Introduction to the Theory of Algebraic Surfaces. IV, 100 pages. 1969. DM 8,– / $ 2.20

Vol. 84: H. Lüneburg, Transitive Erweiterungen endlicher Permutationsgruppen. IV, 119 Seiten. 1969. DM 10,– / $ 2.80

Vol. 85: P. Cartier et D. Foata, Problèmes combinatoires de commutation et réarrangements. IV, 88 pages. 1969. DM 8,– / $ 2.20

Vol. 86: Category Theory, Homology Theory and their Applications I. Edited by P. Hilton. VI, 216 pages. 1969. DM 16,– / $ 4.40

Vol. 87: M. Tierney, Categorical Constructions in Stable Homotopy Theory. IV, 65 pages. 1969. DM 6,– / $ 1.70

Vol. 88: Séminaire de Probabilités III. IV, 229 pages. 1969. DM 18,– / $ 5.00

Vol. 89: Probability and Information Theory. Edited by M. Behara, K. Krickeberg and J. Wolfowitz. IV, 256 pages. 1969. DM 18,– / $ 5.00

Vol. 90: N. P. Bhatia and O. Hajek, Local Semi-Dynamical Systems. II, 157 pages. 1969. DM 14,– / $ 3.90

Vol. 91: N. N. Janenko, Die Zwischenschrittmethode zur Lösung mehrdimensionaler Probleme der mathematischen Physik. VIII, 194 Seiten. 1969. DM 16,80 / $ 4.70

Vol. 92: Category Theory, Homology Theory and their Applications II. Edited by P. Hilton. V, 308 pages. 1969. DM 20,– / $ 5.50

Vol. 93: K. R. Parthasarathy, Multipliers on Locally Compact Groups. III, 54 pages. 1969. DM 5,60 / $ 1.60

Vol. 94: M. Machover and J. Hirschfeld, Lectures on Non-Standard Analysis. VI, 79 pages. 1969. DM 6,– / $ 1.70

Vol. 95: A. S. Troelstra, Principles of Intuitionism. II, 111 pages. 1969. DM 10,– / $ 2.80

Vol. 96: H.-B. Brinkmann und D. Puppe, Abelsche und exakte Kategorien, Korrespondenzen. V, 141 Seiten. 1969. DM 10,– / $ 2.80

Vol. 97: S. O. Chase and M. E. Sweedler, Hopf Algebras and Galois theory. II, 133 pages. 1969. DM 10,– / $ 2.80

Vol. 98: M. Heins, Hardy Classes on Riemann Surfaces. III, 106 pages. 1969. DM 10,– / $ 2.80

Vol. 99: Category Theory, Homology Theory and their Applications III. Edited by P. Hilton. IV, 489 pages. 1969. DM 24,– / $ 6.60

Vol. 100: M. Artin and B. Mazur, Etale Homotopy. II, 196 Seiten. 1969. DM 12,– / $ 3.30

Vol. 101: G. P. Szegö et G. Treccani, Semigruppi di Trasformazioni Multivoche. VI, 177 pages. 1969. DM 14,– / $ 3.90

Vol. 102: F. Stummel, Rand- und Eigenwertaufgaben in Sobolewschen Räumen. VIII, 386 Seiten. 1969. DM 20,– / $ 5.50

Vol. 103: Lectures in Modern Analysis and Applications I. Edited by C. T. Taam. VII, 162 pages. 1969. DM 12,– / $ 3.30

Vol. 104: G. H. Pimbley, Jr., Eigenfunction Branches of Nonlinear Operators and their Bifurcations. II, 128 pages. 1969. DM 10,– / $ 2.80

Vol. 105: R. Larsen, The Multiplier Problem. VII, 284 pages. 1969. DM 18,– / $ 5.00

Vol. 106: Reports of the Midwest Category Seminar III. Edited by S. Mac Lane. III, 247 pages. 1969. DM 16,– / $ 4.40

Vol. 107: A. Peyerimhoff, Lectures on Summability. III, 111 pages. 1969. DM 8,– / $ 2.20

Vol. 108: Algebraic K-Theory and its Geometric Applications. Edited by R. M. F. Moss and C. B. Thomas. IV, 86 pages. 1969. DM 6,– / $ 1.70

Vol. 109: Conference on the Numerical Solution of Differential E tions. Edited by J. Ll. Morris. VI, 275 pages. 1969. DM 18,– / $

Vol. 110: The Many Facets of Graph Theory. Edited by G. Char and S. F. Kapoor. VIII, 290 pages. 1969. DM 18,– / $ 5.00

Vol. 111: K. H. Mayer, Relationen zwischen charakteristischen Za III, 99 Seiten. 1969. DM 8,– / $ 2.20

Vol. 112: Colloquium on Methods of Optimization. Edite N. N. Moiseev. IV, 293 pages. 1970. DM 18,– / $ 5.00

Vol. 113: R. Wille, Kongruenzklassengeometrien. III, 99 Seiten. DM 8,– / $ 2.20

Vol. 114: H. Jacquet and R. P. Langlands, Automorphic Forms on G VII, 548 pages. 1970. DM 24,– / $ 6.60

Vol. 115: K. H. Roggenkamp and V. Huber-Dyson, Lattices over Or XIX, 290 pages. 1970. DM 18,– / $ 5.00

Vol. 116: Séminaire Pierre Lelong (Analyse) Année 1969. IV, 195 p 1970. DM 14,– / $ 3.90

Vol. 117: Y. Meyer, Nombres de Pisot, Nombres de Salem et Ar Harmonique. 63 pages. 1970. DM 6.– / $ 1.70

Vol. 118: Proceedings of the 15th Scandinavian Congress, Oslo Edited by K. E. Aubert and W. Ljunggren. IV, 162 pages. 1970. DM 12,– / $ 3.30

Vol. 119: M. Raynaud, Faisceaux amples sur les schémas en gr et les espaces homogènes. III, 219 pages. 1970. DM 14,– / $ 3.

Vol. 120: D. Siefkes, Büchi's Monadic Second Order Succ Arithmetic. XII, 130 Seiten. 1970. DM 12,– / $ 3.30

Vol. 121: H. S. Bear, Lectures on Gleason Parts. III, 47 pages. DM 6,–/$ 1.70

Vol. 122: H. Zieschang, E. Vogt und H.-D. Coldewey, Fläche ebene diskontinuierliche Gruppen. VIII, 203 Seiten. 1970. DM $ 4.40

Vol. 123: A. V. Jategaonkar, Left Principal Ideal Rings. VI, 145 1970. DM 12,– / $ 3.30

Vol. 124: Séminare de Probabilités IV. Edited by P. A. Meyer. I pages. 1970. DM 20,– / $ 5.50

Vol. 125: Symposium on Automatic Demonstration. V, 310 pages DM 20,– / $ 5.50

Vol. 126: P. Schapira, Théorie des Hyperfonctions. XI, 157 pages DM 14,– / $ 3.90

Vol. 127: I. Stewart, Lie Algebras. IV, 97 pages. 1970. DM 10,– /

Vol. 128: M. Takesaki, Tomita's Theory of Modular Hilbert Al and its Applications. II, 123 pages. 1970. DM 10,– / $ 2.80

Vol. 129: K. H. Hofmann, The Duality of Compact Semigrou C*- Bigebras. XII, 142 pages. 1970. DM 14,– / $ 3.90

Vol. 130: F. Lorenz, Quadratische Formen über Körpern. II, 77 1970. DM 8,– / $ 2.20

Vol. 131: A Borel et al., Seminar on Algebraic Groups and R Finite Groups. VII, 321 pages. 1970. DM 22,– / $ 6.10

Vol. 132: Symposium on Optimization. III, 348 pages. 1970. DM $ 6.10

Vol. 133: F. Topsøe, Topology and Measure. XIV, 79 pages. DM 8,– / $ 2.20

Vol. 134: L. Smith, Lectures on the Eilenberg-Moore Spectral Seq VII, 142 pages. 1970. DM 14,– / $ 3.90

Vol. 135: W. Stoll, Value Distribution of Holomorphic Maps into pact Complex Manifolds. II, 267 pages. 1970. DM 18,– / $

Vol. 136: M. Karoubi et al., Séminaire Heidelberg-Saarbrücker buorg sur la K-Théorie. IV, 264 pages. 1970. DM 18,– / $ 5.00

Vol. 137: Reports of the Midwest Category Seminar IV. Ed S. MacLane. III, 139 pages. 1970. DM 12,– / $ 3.30

Vol. 138: D. Foata et M. Schützenberger, Théorie Géométriq Polynômes Eulériens. V, 94 pages. 1970. DM 10,– / $ 2.80

Vol. 139: A. Badrikian, Séminaire sur les Fonctions Aléatoires aires et les Mesures Cylindriques. VII, 221 pages. 1970. DM $ 5.00

Vol. 140: Lectures in Modern Analysis and Applications II. Ed C. T. Taam. VI, 119 pages. 1970. DM 10,– / $ 2.80

Vol. 141: G. Jameson, Ordered Linear Spaces. XV, 194 pages DM 16,– / $ 4.40

Vol. 142: K. W. Roggenkamp, Lattices over Orders II. V, 388 1970. DM 22,– / $ 6.10

Vol. 143: K. W. Gruenberg, Cohomological Topics in Group T XIV, 275 pages. 1970. DM 20,– / $ 5.50

44: Seminar on Differential Equations and Dynamical Systems, ted by J. A. Yorke. VIII, 268 pages. 1970. DM 20,– / $ 5.50

45: E. J. Dubuc, Kan Extensions in Enriched Category Theory. 73 pages. 1970. DM 16,– / $ 4.40

46: A. B. Altman and S. Kleiman, Introduction to Grothendieck y Theory. II, 192 pages. 1970. DM 18,– / $ 5.00

47: D. E. Dobbs, Cech Cohomological Dimensions for Comve Rings. VI, 176 pages. 1970. DM 16,– / $ 4.40

48: R. Azencott, Espaces de Poisson des Groupes Localement acts. IX, 141 pages. 1970. DM 14,– / $ 3.90

49: R. G. Swan and E. G. Evans, K-Theory of Finite Groups and s. IV, 237 pages. 1970. DM 20,– / $ 5.50

50: Heyer, Dualität lokalkompakter Gruppen. XIII, 372 Seiten. DM 20,– / $ 5.50

51: M. Demazure et A. Grothendieck, Schémas en Groupes I. 3). XV, 562 pages. 1970. DM 24,– / $ 6.60

52: M. Demazure et A. Grothendieck, Schémas en Groupes II. 3). IX, 654 pages. 1970. DM 24,– / $ 6.60

53: M. Demazure et A. Grothendieck, Schémas en Groupes III. 3). VIII, 529 pages. 1970. DM 24,– / $ 6.60

54: A. Lascoux et M. Berger, Variétés Kähleriennes Compactes. pages. 1970. DM 8,– / $ 2.20

55: J. J. Horváth, Several Complex Variables, I, Maryland 1970. IV. ges. 1970. DM 18,– / $ 5.00

56: R. Hartshorne, Ample Subvarieties of Algebraic Varieties. 56 pages. 1970. DM 20,– / $ 5.50

57: T. tom Dieck, K. H. Kamps und D. Puppe, Homotopietheorie. 5 Seiten. 1970. DM 20,– / $ 5.50

8: T. G. Ostrom, Finite Translation Planes. IV. 112 pages. 1970. ,– / $ 2.80

59: R. Ansorge und R. Hass. Konvergenz von Differenzenverfür lineare und nichtlineare Anfangswertaufgaben. VIII, 145 1970. DM 14,– / $ 3.90

0: L. Sucheston, Constributions to Ergodic Theory and Proba-VII, 277 pages. 1970. DM 20,– / $ 5.50

61: J. Stasheff, H-Spaces from a Homotopy Point of View. pages. 1970. DM 10,– / $ 2.80

2: Harish-Chandra and van Dijk, Harmonic Analysis on Reducadic Groups. IV, 125 pages. 1970. DM 12,– / $ 3.30

63: P. Deligne, Equations Différentielles à Points Singuliers ers. III, 133 pages. 1970. DM 12,– / $ 3.30

4: J. P. Ferrier, Seminaire sur les Algebres Complètes. II, 69 pa-970. DM 8,– / $ 2.20

5: J. M. Cohen, Stable Homotopy. V, 194 pages. 1970. DM 16,– /

6: A. J. Silberger, PGL_2 over the p-adics: its Representations, cal Functions, and Fourier Analysis. VII, 202 pages. 1970. ,– / $ 5.00

7: Lavrentiev, Romanov and Vasiliev, Multidimensional Inverse ms for Differential Equations. V, 59 pages. 1970. DM 10,– / $ 2.80

8: F. P. Peterson, The Steenrod Algebra and its Applications: ference to Celebrate N. E. Steenrod's Sixtieth Birthday. VII, ges. 1970. DM 22,– / $ 6.10

9: M. Raynaud, Anneaux Locaux Henséliens. V, 129 pages. 1970. – / $ 3.30

0: Lectures in Modern Analysis and Applications III. Edited by aam. VI, 213 pages. 1970. DM 18,– / $ 5.00.

1: Set-Valued Mappings, Selections and Topological Properties dited by W. M. Fleischman. X, 110 pages. 1970. DM 12,– / $ 3.30

2: Y.-T. Siu and G. Trautmann, Gap-Sheaves and Extension erent Analytic Subsheaves. V, 172 pages. 1971. DM 16,– / $ 4.40

3: J. N. Mordeson and B. Vinograde, Structure of Arbitrary Inseparable Extension Fields. IV, 138 pages. 1970. DM 14,– /

4: B. Iversen, Linear Determinants with Applications to the Scheme of a Family of Algebraic Curves. VI, 69 pages. 1970. – / $ 2.20.

5: M. Brelot, On Topologies and Boundaries in Potential Theory. pages. 1971. DM 18,– / $ 5.00

Vol. 176: H. Popp, Fundamentalgruppen algebraischer Mannigfaltigkeiten. IV, 154 Seiten. 1970. DM 16,– / $ 4.40

Vol. 177: J. Lambek, Torsion Theories, Additive Semanics and Rings of Quotients. VI, 94 pages. 1971. DM 10,– / $ 2.80

Vol. 178: Th. Bröcker und T. tom Dieck, Kobordismentheorie. XVI, 191 Seiten. 1970. DM 18,– / $ 5.00

Vol. 179: Seminaire Bourbaki – vol. 1968/69. Exposés 347-363. IV. 295 pages. 1971. DM 22,– / $ 6.10

Vol. 180: Séminaire Bourbaki – vol. 1969/70. Exposés 364-381. IV, 310 pages. 1971. DM 22,– / $ 6.10

Vol. 181: F. DeMeyer and E. Ingraham, Separable Algebras over Commutative Rings. V, 157 pages. 1971. DM 16.– / $ 4.40

Lecture Notes in Physics

Bisher erschienen/Already published

Vol. 1: J. C. Erdmann, Wärmeleitung in Kristallen, theoretische Grundlagen und fortgeschrittene experimentelle Methoden. II, 283 Seiten. 1969. DM 20,– / $ 5.50

Vol. 2: K. Hepp, Théorie de la renormalisation. III, 215 pages. 1969. DM 18,– / $ 5.00

Vol. 3: A. Martin, Scattering Theory: Unitarity, Analytic and Crossing. IV, 125 pages. 1969. DM 14,– / $ 3.90

Vol. 4: G. Ludwig, Deutung des Begriffs physikalische Theorie und axiomatische Grundlegung der Hilbertraumstruktur der Quantenmechanik durch Hauptsätze des Messens. XI, 469 Seiten. 1970. DM 28,– / $ 7.70

Vol. 5: M. Schaaf, The Reduction of the Product of Two Irreducible Unitary Representations of the Proper Orthochronous Quantummechanical Poincaré Group. IV, 120 pages. 1970. DM 14,– / $ 3.90

Vol. 6: Group Representations in Mathematics and Physics. Edited by V. Bargmann. V, 340 pages. 1970. DM 24,– / $ 6.60